A BRIEF HISTORY OF MATHEMATICS
FOR CURIOUS MINDS

A BRIEF HISTORY OF MATHEMATICS FOR CURIOUS MINDS

Krzysztof R. Apt

Centrum Wiskunde & Informatica (CWI), Amsterdam, The Netherlands

MIMUW, University of Warsaw, Poland

World Scientific

NEW JERSEY · LONDON · SINGAPORE · BEIJING · SHANGHAI · HONG KONG · TAIPEI · CHENNAI · TOKYO

Published by

World Scientific Publishing Co. Pte. Ltd.

5 Toh Tuck Link, Singapore 596224

USA office: 27 Warren Street, Suite 401-402, Hackensack, NJ 07601

UK office: 57 Shelton Street, Covent Garden, London WC2H 9HE

Library of Congress Cataloging-in-Publication Data

Names: Apt, Krzysztof R., 1949–   author.

Title: A brief history of mathematics for curious minds / Krzysztof R. Apt,
    Centrum Wiskunde & Informatica (CWI), Amsterdam, The Netherlands, MIMUW,
    University of Warsaw, Poland.

Description: New Jersey : World Scientific, [2024] | Includes bibliographical references and index.

Identifiers: LCCN 2023034145 | ISBN 9789811280443 (hardcover) |
    ISBN 9789811281495 (paperback) | ISBN 9789811280450 (ebook for institutions) |
    ISBN 9789811280467 (ebook for individuals)

Subjects: LCSH: Mathematics--History--Popular works.

Classification: LCC QA21 .A65 2024 | DDC 510.9--dc23/eng/20231016

LC record available at https://lccn.loc.gov/2023034145

British Library Cataloguing-in-Publication Data

A catalogue record for this book is available from the British Library.

For any available supplementary material, please visit
https://www.worldscientific.com/worldscibooks/10.1142/13518#t=suppl

To Ezra Apt, wishing him a happy life

Preface

Mathematics is part of our cultural heritage, just as how philosophy, art, history, and great works of literature are. Yet, even with a lamp in broad daylight, one can hardly find an intellectual who would be able to say anything meaningful about the achievements of mathematicians, let alone a well-educated person. Most history textbooks devote hardly any attention to mathematics and mathematicians.[1]

I would like to offer here a short and accessible account of the history of mathematics, written for an intelligent layman, hoping that it will allow him or her to better appreciate its beauty, relevance, and place in history. And, as you will see, the cast of characters is at least as fascinating as that of philosophers or writers.

It is natural to assume that mathematics originated because of the need for *counting*. More ambitious tasks, such as the determination of distances and areas, and also of taxes, required *calculating*. Soon after or perhaps at the same time, mathematics was used to deal with more challenging problems posed by astronomy and physics. During the Renaissance, mathematics was already applied in optics, mechanics, geography, cartography, and even in painting (in the concept of perspective). As time went on, more and more uses for mathematics were found. For instance, as early as 1762, insurance companies started to use mathematics in determining their premiums.[2]

Since then, mathematics has contributed to progress in almost all sciences, including psychology (statistical analysis), sociology (network analysis), medicine (e.g., geometry and calculus are used in CT scan technology and in virology), chemistry (e.g., graph theory), crystallography (algebra), economics (e.g., calculus and game theory), political science (voting theory), meteorology (dynamical systems), biology (e.g., calculus), computer science (e.g., mathematical logic, combinatorics), linguistics (e.g., formal grammars), geosciences

(Fourier analysis and wavelet analysis), and even history (radiocarbon dating uses logarithms). Some mathematicians won for their works Nobel Prizes in physics, economics, and medicine.

There are two competing ways of telling the story of mathematics. One is through a chronological account, introducing mathematicians who successively enter the hall of fame. The other is to discuss the areas of mathematics and the notions they focus on. Ideally, to reconcile both approaches a book should not have the customary linear form but instead consist of a grid formed by the entries (mathematician/area) combining both approaches. But then, the outcome would be a handbook instead of a book.

I chose the chronological approach here, though, in my enthusiasm, I occasionally jump forward, pursuing the impact of an idea. In doing so, I may be guilty of too many digressions, but hopefully, they will allow the reader to appreciate the boldness of some of the early achievements.

It is not easy to provide an accessible account of the history of mathematics, for at least two reasons. One is that mathematics is a vast subject that is continuously growing. This makes it difficult to write about it in a competent way. The other is that starting from the 16th century, we enter a period in which mathematicians introduce more and more technical concepts. In jargon used by mathematicians, 'technical' usually means 'difficult to explain'.

Such concepts, to mention a few, include functions, relations, operations, real (numbers), complex (numbers), groups, rings, fields, filters, kernels, lattices, grids, spaces, modules, orders, matrices, faces, graphs, trees, interpretations, transformations, and models. All these concepts, just like adjectives, e.g., ideal, rational, irrational, open, closed, dense, sparse, cardinal, and discrete, have completely different meanings outside of mathematics. In particular, the mathematical concept of commuting has nothing to do with traveling and trees in mathematics may be infinite and usually grow downwards.

Other concepts such as line, induction, dimension, vector, probability, tautology, continuous (change), or limit seem to mean the same both within and outside of mathematics, but mathematics offers precise definitions of them, which, incidentally, have often been born out of centuries of long struggles, discussions, and occasional animosities.

A case apart is the concept of infinity, which has been troubling mathematicians (and philosophers) since the time of the Greeks. In 1784, the Berlin Academy offered a prize for a "clear and precise theory of what is the infinite in mathematics". It was not satisfied with any of the entries.[3] The problems associated with infinity continued to bother mathematicians of the 19th century and prompted them to substantially redefine the basic notions of an important branch of mathematics. Yet, in 1925, David Hilbert, a German mathematician, wrote: "If we pay close attention, we find that the literature of mathematics is replete with absurdities and inanities, which can usually be blamed on the infinite." He further wrote (italics in the original): "[...] the definitive clarification of the *nature of the infinite* has become necessary, not merely for the special interests of the individual sciences but for the *honor of the human understanding itself*."[4] Apparently, this honor has not been saved yet—in 2013, Scientific American ran an article *Dispute over infinity divides mathematicians* that discusses two competing theories of infinite sets over which logicians currently debate.[5]

Yet another difficulty in presenting the history of mathematics to an educated layman is that the driving force behind mathematics is *generalization*. Over centuries, mathematicians have generalized almost all possible concepts to the extent they even occasionally question their usefulness.

The simplest example of a generalization is the concept of a *number*. The Babylonians accepted numbers such as $1, 2, 3 \ldots$, as well as *unit fractions* (for example, $\frac{1}{6}$). The Greeks studied special numbers, called *prime numbers* and also begrudgingly agreed to admit *irrational numbers*, such as $\sqrt{2}$. The Chinese and Indians introduced negative numbers, the usual fractions, and enriched us with the number zero. This way we obtained *integers*— $\ldots, -3, -2, -1, 0, 1, 2, 3, \ldots$,—and real numbers, (*reals*, for short), e.g., $-5\frac{1}{3}, -\sqrt{3}, 0, 1\frac{1}{2}$, or $\sqrt{2}$.

However, in the 16th century, out of the work done on solving third-degree equations (for instance, $x^3 - 15x - 4 = 0$), there emerged *complex numbers*, a concept that goes beyond secondary school mathematics. Then we also have *transcendental numbers*, p-*adic numbers*, and *algebraic numbers*. To top it off Georg Cantor introduced *infinite numbers* (called *cardinals*, a shorthand for a *cardinal number*) in the late 19th century. So in the end, the word 'number' is a little bit ambiguous in mathematics, to say the least.[6]

In what follows by a *number*, I shall usually mean 1, 2, 3, etc.; so, what is called in mathematical parlance a *natural number*. An excellent way to lose the reader is to delve into the notion that a natural number should not be taken for granted. So of course I shall not do that.

Any informal account of the history of mathematics is easily prone to criticism. Mathematicians will complain that several important areas of mathematics and mathematicians are not mentioned, while a 'general reader' will soon be lost if one assumes from him or her knowledge of mathematics at the secondary school level. To solve this dilemma and to keep the account informal, I moved all technical comments and references to footnotes. Further, 32 appendices provide more detailed information and selected proofs, some less known but all at an elementary level. Finally, at the end, I list some novels and films about mathematicians and recommend various books on the history of mathematics that can be of interest to a 'general reader'.

Acknowledgments

I would like to thank José María Almira who read the initial manuscript and provided several useful comments that led to various improvements in the presentation. I also profited from several useful suggestions made by Alma Apt, Jan Heering, Helena Stockmann, Nick Trefethen, Ronald de Wolf, the late Maarten van Emden, and the three anonymous referees.

This book could not have been written without access to the wonderful CWI library. I am most grateful to its staff, in particular its director Vera Sarkol, and also Rob van Rooijen, for having ordered several books on the history of mathematics that were highly relevant. Further, I would like to thank Magdalena Kycler and Piotr Sitek for their expert production of several drawings using the Ti*k*z package. My special appreciation goes to Ms. Lai Fun Kwong of World Scientific for a most efficient and smooth cooperation, and to Gregory Lee for his remarkably thorough desk copy editing.

This book is set using the fbb LATEX package that provides a free Bembo-like font designed by Michael Sharpe.

All proceeds from this book will be donated to Amnesty International.

Notes

[1] Notable exceptions are D.J. Boorstin, *The Discoverers: A History of Man's Search to Know His World and Himself*, Vintage, 1985, and P. Watson, *Ideas: A History of Thought and Invention, from Fire to Freud*, Weidenfeld & Nicholson, 2005.

[2] K. Devlin, *The Unfinished Game: Pascal, Fermat, and the Seventeenth-Century Letter that Made the World Modern*, Basic Books, 2008.

[3] D. Bressoud, *A Radical Approach to Real Analysis*, The Mathematical Association of America, p. 52, 2007.

[4] D. Hilbert, On the Infinite, in: J. van Heijenoort, *From Frege to Gödel*, Harvard University Press, pp. 367–392, 1967.

[5] N. Wolchover, Dispute over infinity divides mathematicians, *Scientific American*, 3 December 2013, https://www.scientificamerican.com/article/infinity-logic-law/. Originally published in *Quanta Magazine*. For a recent book on the subject, see I. Stewart, *Infinity: a Very Short Introduction*, Oxford University Press, 2017.

[6] Consider this remark: "Today it is no longer that easy to decide what counts as a 'number'." from F.Q. Gouvêa, From Numbers to Number Systems, in: T. Gowers, I. Leader, and J. Barrow-Green (eds.), *The Princeton Companion to Mathematics*, Princeton University Press, pp. 77–82, 2008.

Contents

Chapter 1

From the Beginnings to 6th Century BCE

Tally marks It makes sense to assume that the first instance of mathematical activity was *counting*. There is evidence that even birds are able to distinguish small numbers, say four from five. Early humans could do better, of course.

In the 1970s, the so-called *Lebombo Bone*, a baboon's leg bone with 29 tally marks, was discovered in Swaziland. One speculates that this number has to do with the fact that it takes the Moon 29 days to circle the Earth. In fact, the first civilizations used a lunar calendar. The Lebombo Bone is the oldest known mathematical object, estimated to be at least 43,000 years old. Its age exceeds that of an approximately 30,000-year-old wolf bone with 55 tally marks, excavated in 1937 in the Czech Republic. A small attempt at organizing the resulting information was made as the marks were divided into groups of 5.

Another bone with tally marks, approximately 20,000 years old, became, in some sense, a bone of contention. Called the *Ishango bone*, it was discovered in 1950 in what is now the Democratic Republic of the Congo. It contains 168 notches arranged in different patterns in three columns. Many theories have been put forward to explain the meaning of the used arrangements. In particular, one claims that it records different phases of the Moon, while another claims that it constitutes a precursor of writing. Determining which theory is right is difficult, given the lack of additional evidence.[1]

Still, in modern terminology, these bones represent counting in *unary notation* as they rely on a single symbol, a mark. Progress was achieved by inventing notations that could rely on more symbols.

Babylonians Two ancient cultures substantially contributed to developing mathematics. One of them was the Babylonian civilization, named after the city of Babylon in Iraq. It emerged around 2000 BCE. The principal source of information on it is the collection of clay tablets. After the scribes wrote on them (from left to right) the tablets were dried or baked. This process ensured that several survived until today. The tablets of greatest mathematical interest were written in the period 1800–1600 BCE, in the times when the extensive law code of Hammurabi was established.

The Babylonians developed a number system that had 59 different symbols for the numbers 1 to 59. In modern terminology, this is a system with base 60 with 0 missing. The motivation for such a peculiar choice is not clear. It has been suggested that it arose as a merger of two other number systems.

1	𒁹	11	𒌋𒁹	21	𒎙𒁹	31	𒌍𒁹	41	𒐏𒁹	51	𒐐𒁹
2	𒈫	12	𒌋𒈫	22	𒎙𒈫	32	𒌍𒈫	42	𒐏𒈫	52	𒐐𒈫
3	𒐈	13	𒌋𒐈	23	𒎙𒐈	33	𒌍𒐈	43	𒐏𒐈	53	𒐐𒐈
4	𒐉	14	𒌋𒐉	24	𒎙𒐉	34	𒌍𒐉	44	𒐏𒐉	54	𒐐𒐉
5	𒐊	15	𒌋𒐊	25	𒎙𒐊	35	𒌍𒐊	45	𒐏𒐊	55	𒐐𒐊
6	𒐋	16	𒌋𒐋	26	𒎙𒐋	36	𒌍𒐋	46	𒐏𒐋	56	𒐐𒐋
7	𒐌	17	𒌋𒐌	27	𒎙𒐌	37	𒌍𒐌	47	𒐏𒐌	57	𒐐𒐌
8	𒐍	18	𒌋𒐍	28	𒎙𒐍	38	𒌍𒐍	48	𒐏𒐍	58	𒐐𒐍
9	𒐎	19	𒌋𒐎	29	𒎙𒐎	39	𒌍𒐎	49	𒐏𒐎	59	𒐐𒐎
10	𒌋	20	𒎙	30	𒌍	40	𒐏	50	𒐐		

Babylonian numerals.

The above listing of Babylonian numerals shows that the symbols used are not independent: those representing numbers larger than 10 are pairs of two symbols. For example, the symbol for 37 is 𒌍𒐌, which is a combination of the symbols for 30 and 7: 𒌍 and 𒐌 .

This way of counting with base 60 has survived until today, as seen in our division of hours into 60 minutes and minutes into 60 seconds. Also, our division of a day into 24 hours and a circle into 360 degrees goes back to the Babylonians.

The Babylonian number system was *positional*, which means that a symbol represents a *different* value depending on the position it stands. Larger numbers were written in this positional notation using the above 59 numerals, just like we do in our decimal system. Spacing was used instead of the digit zero, but this, of course, could cause ambiguities.[2]

In 1854 two tablets from about 2000 BCE were found that contained squares and cubes of numbers up to 59 and 32, respectively. This shows how impressive the calculating skills of the Babylonians were. To divide the numbers, they realized that $\frac{a}{b} = a \cdot \frac{1}{b}$, which brought them the idea of constructing tables of the unit fractions (i.e., fractions of the form $\frac{1}{b}$) and recording them in their base 60 arithmetic.[3]

Some complications arose with fractions like $\frac{1}{7}$, which could not be written in such a form. A scribe was aware of the problem, admitting that "an approximation is given since 7 does not divide".[4] Of course, a similar problem arises in the decimal notation.

The Babylonians used mathematics for accounting, trade, time recording, astronomical observations, and surveying purposes. Preserved clay tablets record solutions of simple mathematical problems that involve one or two variables that led to what we call *linear equations*. Here is an example problem from a text found during excavations in Iraq in 1949; *qa* is a weight unit.[5]

> If somebody asks you thus: If I add to the two-thirds of my two-thirds a hundred qa of barley, the original quantity is summed up. How much is the original quantity?[6]

However, no general methods were proposed, and each problem was solved individually. Another tablet showed that the Babylonians determined the value of π, the ratio of the circumference of the circle and its diameter, as $3\frac{1}{8}$, i.e., 3.125. Other discovered tablets revealed that the Babylonians were also able to solve specific *quadratic* (i.e., second-degree) equations and even gave a try at some specific *third-degree* equations. Here is an example of a typical problem that led to a quadratic equation:[7]

> The length of a rectangle exceeds its width by 7. Its area is 60. Find its length and width.[8]

Another important source confirming the sophistication of Babylonian mathematics is *Plimpton 322*, a small clay tablet measuring 12.7 centimeters by 8.8 centimeters (so it would fit into an A6-sized envelope). It was purchased in 1922 for \$10 by an American publisher, George Arthur Plimpton, from an antiquarian. It comes from Iraq and is estimated to have been written around 1800 BCE. In 1945, two mathematicians found out that the tablet provided information about 15 so-called Pythagorean triples, i.e., the numbers a, b, and c, such that $a^2 + b^2 = c^2$ (for instance 3, 4, and 5, as $3^2 + 4^2 = 5^2$).[9]

Egyptians Essentially in parallel with the Babylonians, the Egyptian civilization developed. The Egyptians recorded their writing on papyri using hieroglyphs. The usual writing direction was from right to left.

The Plimpton 322 tablet.[10]

Our knowledge of Egyptian mathematics stems from two manuscripts. The first is the *Rhind Mathematical Papyrus* from approximately 1650 BCE. It contains material copied from an original that is about 200 years older. The papyrus was purchased in 1858 by a Scottish antiquary, Henry Rhind, and acquired after his death by the British Museum in London. It is about 536 centimeters by 32 centimeters, but with some parts missing. By a remarkable coincidence, some useful missing fragments were found half a century later in the deposits of a New York museum.

The papyrus is probably today's equivalent of a mathematics textbook. It includes 84 problems concerned with divisions, multiplication, and handling of fractions. An example of a problem tackled is how to divide 6 loaves among 10 men. The answer is $\frac{1}{2} + \frac{1}{10}$, as the Egyptians, like Babylonians, used unit fractions (with the single exception of $\frac{2}{3}$) exclusively. Also, more advanced problems were dealt with; these amounted to solving linear equations in one variable. For instance, another problem calls for finding the size of a scoop that requires $3\frac{1}{3}$ trips to a well to fill one *heqat* (an ancient Egyptian unit of volume). In modern terminology, this is simply the equation $3\frac{1}{3}x = 1$, with the answer $x = \frac{3}{10}$.

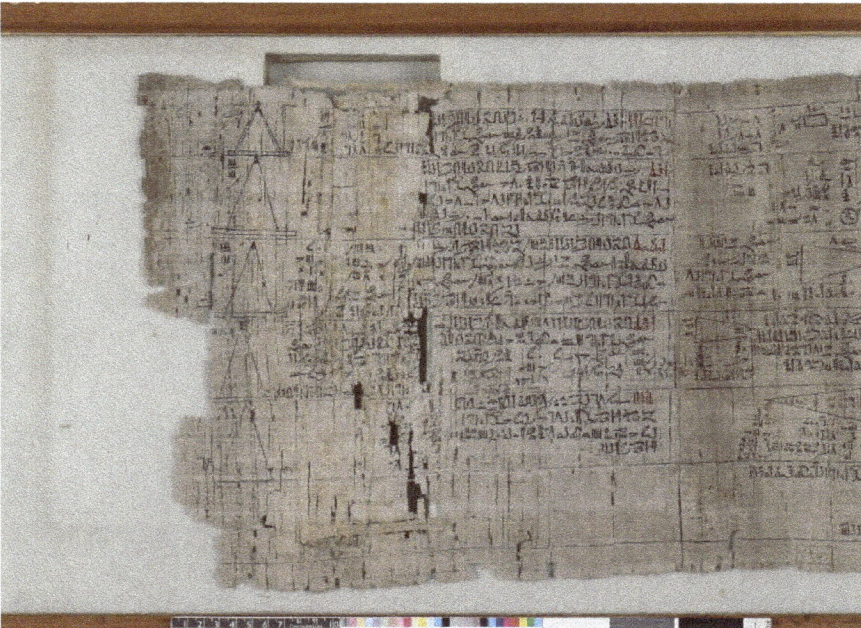

The Rhind Mathematical Papyrus.[11]

The papyrus also contains geometry problems. In particular, it provided a method of calculating π, the value of which was determined as $\frac{256}{81}$, i.e., about 3.16. One of the problems contains the following puzzle:

> Seven houses have seven cats that each eats seven mice that each eats seven grains of barley. Each barley grain would have produced seven heqats of grain.

The task is to determine how many things are described. This is clearly an exercise in adding up consecutive powers: the answer is $7 + 7^2 + 7^3 + 7^4 + 7^5 = 19,607$.

So, this is a precursor to the following famous English riddle that dates back to the 18th century (though the answer is here: just one):

> As I was going to St. Ives,
> I met a man with seven wives,
> Each wife had seven sacks,
> Each sack had seven cats,
> Each cat had seven kits:

Kits, cats, sacks, and wives,
How many were there going to St. Ives?

The second manuscript is the *Moscow Mathematical Papyrus*, purchased in Egypt by a Russian Egyptologist around 1892. It is now part of a collection of the Pushkin State Museum of Fine Arts in Moscow, hence its name. The papyrus is about 5.5 meters long and between 4 and 7 centimeters wide and is approximately from 1850 BCE.

It contains 25 problems concerned with arithmetic and the computation of areas and volumes. In particular, it provides a correct formula for computing the volume of a truncated pyramid, like a pyramid 'under construction' with the top part missing. It is the most remarkable achievement of Egyptian mathematics we know of.

To represent numbers, the Egyptians employed a system that used different symbols for each power of 10, all the way up to 1,000,000:

Numbers were then written by repeating these symbols the appropriate number of times, sometimes displayed in more rows. Armed with this knowledge, the reader can easily decipher information written on some old artifacts.

A fragment of the macehead of King Narmer.[12]

For example, in 1898, a macehead from about 3100 BCE was found in Egypt that belonged to King Narmer. It is on display in the Ashmolean Museum in Oxford. The above representation of its fragment shows the number 400,000 (the lower part under the sign of an ox) written as

(in two columns), the number 1,422,000 (under and next to the sign of a goat) written as

and to the right, the number 120,000 (under the sign of a prisoner (a sitting man with his arms tied behind his back)) written as

This macehead shows that the Egyptians were already using their number system more than 5,000 years ago.

Finally, to represent fractions, a system eventually emerged according to which the unit fraction $\frac{1}{n}$ was written by putting the symbol ⬭ over n. For example, $\frac{1}{15}$ was written as

The Nile and its floodings were crucial for the Egyptians. Consequently, they developed a solar calendar early on to predict them. It eventually replaced a lunar calendar. In their calendar, a year had 365 days and was divided into 12 months of 30 days, with 5 additional days at the end. Due to the lack of leap years, the calendar year drifted farther and farther away from the solar year and after 1,460 years, made a full circle, which must have happened more than once. This problem was eventually addressed by Julius Caesar during the 1st century BCE.

Notes

[1]G.G. Joseph, *The Crest of the Peacock: Non-European Roots of Mathematics*, Princeton University Press, 3rd edition, pp. 33–35, 2011.

[2]For example, 62 (so $1 \cdot 60^1 + 2$) is represented as ⟇ ⟇⟇ , while 3,602 (so $1 \cdot 60^2 + 0 \cdot 60^1 + 2$) is represented by ⟇ ⟇⟇ .

[3]An example may help to understand this idea. The decimal notation uses base 10, so the fraction $\frac{1}{25}$ is written as 0.04, since $\frac{1}{25} = \frac{0}{10} + \frac{4}{10^2}$. In the Babylonians' approach, base 60 is used, so $\frac{1}{25}$ was recorded as the sequence of two numerals ⟇⤳⤶ representing 2 and 24, since $\frac{1}{25} = 2 \cdot \frac{1}{60} + 24 \cdot \frac{1}{60^2}$.

[4]J.J. O'Connor and E.F. Robertson, An overview of Babylonian mathematics, http://www-history.mcs.st-and.ac.uk/HistTopics/Babylonian_mathematics.html, 2000.

[5]G.G. Joseph, op. cit., p. 153.

[6]It leads to a linear equation $\frac{2}{3} \cdot \frac{2}{3} x + 100 = x$, with the solution $x = 180$.

[7]G.G. Joseph, op. cit., p. 154.

[8]It can be formalized by the equations $y = x + 7$ and $xy = 60$. They lead to the equation $x(x + 7) = 60$, which expands to the quadratic equation $x^2 + 7x = 60$, with the solution $x = 5$.

[9]V.J. Katz and K.H. Parshall, *Taming the Unknown: A History of Algebra from Antiquity to the Early Twentieth Century*, Princeton University Press, 2014.

[10]Courtesy: Public Domain, Wikipedia Commons.

[11]Courtesy: Public Domain, Wikipedia Commons.

[12]Courtesy: Public Domain, Wikipedia Commons.

Chapter 2

The Greeks
(From 6th Century BCE to
5th Century CE)

Ancient Greek mathematics covers a period of more than 1,200 years. At that time, there was no division of science into disciplines. So, mathematicians were often also philosophers, astronomers, or engineers, who applied their mathematical knowledge and ideas to these disciplines. But—crucially—they also studied mathematics for its own sake. Their focus was on number theory, geometry, logic, and, occasionally, algebra. This brought them to a study of problems that went far beyond the needs of daily life. Several of their results had no direct use, but, as we shall soon see, blaming mathematicians for their lack of concern with applications can be shortsighted.

Thales and trigonometry We begin our account of Greek mathematics with **Thales of Miletus** (c. 625–545 BCE), a successful oil merchant and one of the wise men of his times. Thales is considered the traditional 'father of philosophy'; he believed that in nature, everything originates from water. He can also be viewed as the 'father of *trigonometry*', the area of mathematics concerned with the study of triangles. Students in secondary schools learn the *Thales' theorem*, which is concerned with a triangle inscribed in a circle. This theorem is discussed in Appendix 1.

Thales was a brilliant astronomer (in particular, he apparently predicted a solar eclipse) and a mathematician who was able to compute the height of the pyramids and the distance of ships at sea from the shore.

Trigonometry gave rise to *triangulation*, a method of creating precise maps by constructing successive triangles. It was invented in the 16th century by a Dutch mathematician, **Gemma Frisius** (1508–1555). Sixty years later, another Dutchman, **Willebrord Snel van Royen** (1580–1626), better known as **Willebrord Snellius**, vastly improved upon it by proposing to measure the angles instead of the sides of the triangles.[1] In spite of this improvement, producing the first precise maps of countries or continents was not a trifling matter. For instance, in the 18th century it took the Italian-French Cassini family some 50 years to produce the first reliable map of France.[2]

Pythagoras and the Pythagorean Theorem

A contemporary of Thales, **Pythagoras of Samos** (c. 570–c. 500 BCE), is credited with the famous *Pythagorean theorem*. It states that in a right-angled triangle, with the side lengths a, b, and c, where c is the length of the hypotenuse, we have $a^2 + b^2 = c^2$. Pythagoras was so delighted with this discovery that he apparently "offered a hundred oxen to the Muses" to thank for the inspiration.[3]

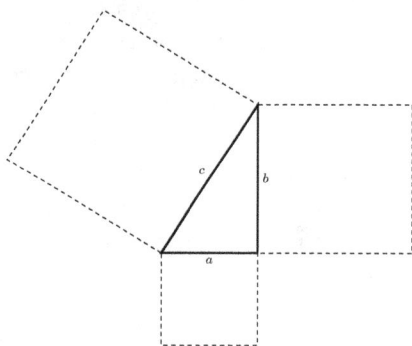

Pythagorean theorem: $a^2 + b^2 = c^2$.

The already discussed Plimpton 322 tablet seems to indicate that specific cases of this result were already known to the Babylonians. Also, the Egyptians knew that a triangle with sides 3, 4, and 5 was right-angled. The proof of the Pythagorean theorem is given in Euclid's *Elements*, a book from around the 3rd century BCE, which I shall discuss shortly. However, Euclid's proof is not so straightforward.

One of the simplest proofs was published in a short article by an American President, James Abram Garfield, who was in office for only 199 days—he died in 1881 after being shot by an assassin. His proof is presented in Appendix 2. Garfield concluded his article with this remark: "We think it something on which the members of both houses can unite without distinction of party."[4]

By now, there are plenty of proofs of the Pythagorean theorem. Already, in 1940, a collection of no less than 370 proofs appeared.[5] New proofs keep appearing. In 2016, a World Bank economist Kaushik Basu published one in a paper succinctly titled *A new and rather long proof of the Pythagoras theorem by way of a proposition on isosceles triangles.*[6]

Pythagoras is considered to be the first person who concluded that the Earth was a sphere, as opposed to a flat circular disc, as was believed among others like Homer. As stated by Bertrand Russell: "Pythagoras [...] was intellectually one of the most important men that ever lived, both when he was wise and when he was unwise. [...] He founded a religion, of which the main tenets were the transmigration of souls and the sinfulness of eating beans."[7] But the school he founded in southern Italy was also concerned with philosophy and mathematics.

One of the symbols used by the school was the *pentagram*, a five-pointed star formed by drawing the diagonals of a *regular pentagon*.[8] The Pythagoreans were fascinated by numbers and believed that the universe could be fully understood in their terms. Apparently, a motto of their school was "All is number".[9]

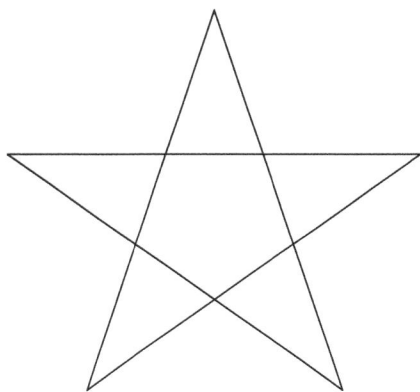

A pentagram.

They realized that the sums of successive odd numbers form consecutive squares, so $1 + 3 = 2^2$, $1 + 3 + 5 = 3^2$, $1 + 3 + 5 + 7 = 4^2$, and so on. The Pythagoreans also invented the concept of a *perfect number*. These are the numbers that are equal to the sum of their proper divisors. For example, 6 is perfect since $1+2+3 = 6$ and so is 28, since $1+2+4+7+14 = 28$. They attributed mystical properties to these perfect numbers. This idea persisted for some time. For example, during the 5th century, St. Augustine claimed that God created the world in 6 days because 6 is a perfect number. Perfect numbers give rise to one of the oldest open problems in mathematics: are all perfect numbers even? Also it is not known whether there are infinitely many perfect numbers.

The Pythagoreans also attached importance to music and discovered that musical intervals between notes could be expressed as numbers. In particular, Pythagoras discovered that halving the length of a string would produce a tone exactly an octave higher when struck or plucked.

π and some other irrational numbers The Pythagoreans were familiar with fractions like $\frac{1}{3}$ or $\frac{2}{5}$, but, to their horror, they discovered that other numbers existed as well. They found, for example, that the diagonal of a square whose sizes are 1 couldn't be expressed as a fraction. Nowadays we write this number as $\sqrt{2}$ and call it the 'square root of 2'.

We happen to encounter this number daily: it is the ratio of the sides of an

A4 paper sheet. The ratio $\sqrt{2}$ is not accidental; it ensures that an A5 paper sheet, obtained by folding an A4 paper sheet in half along its longer edge, has the same proportions as the original A4 paper sheet.

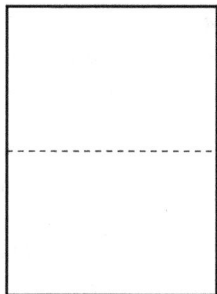

This idea of using the ratio $\sqrt{2}$ is relatively recent: It is due to German physicist Georg Christoph Lichtenberg, who proposed it only in 1786.

Fractions, such as $\frac{2}{5}$ or $\frac{5}{2}$, are nowadays called *rational numbers*, while numbers that cannot be expressed as fractions are called *irrational numbers*. So, $\sqrt{2}$ is an irrational number. A simple proof is given in Appendix 4, where it is also explained why $\sqrt{2}$ is the ratio of the sides of the A4 paper.

Another example of an irrational number is the already discussed π. It is celebrated in many ways, for instance in a 1976 poem by Polish poet Wisława Szymborska, a Nobel Prize laureate, which starts with:

The A4 paper sheet.

> The admirable number pi:
> three point one four one.
> All the following digits are also initial,
> five nine two because it never ends.

In 2009, in a bid for stronger science education, the US House of Representatives approved a resolution designating 14 March 2009 as National Pi Day. (The American way of writing 14 March is 3/14.) Since then, 14 March is internationally celebrated as Pi Day.

Computing the value of π has fascinated mathematicians ever since antiquity and is a recurring theme throughout the history of mathematics, one to which I shall return more than once. The Greeks substantially improved upon the Babylonian and Egyptian estimates. In the 3rd century BCE, Archimedes used a 96-gon (a regular polygon with 96 sides) to approximate the circumference of a circle and concluded that $3\frac{10}{71} < \pi < 3\frac{10}{70}$, i.e., $3.1408\cdots < \pi < 3.1428\ldots$, which gives a correct approximation to two decimal places. A substantially better approximation $\frac{355}{113}$ was found in the 5th century by Chinese mathematician **Zu Chongzhi** and was not improved upon for the next 900 years. This easy-to-remember fraction (think in terms of dividing the sequence of six digits 113355 into two equal parts) yields a correct approximation of π to six decimal places, which is the best one can achieve using a fraction with six digits.

The Dutch-German mathematician **Ludolph van Ceulen** (1540–1610) spent a major part of his life calculating π to 35 decimal places. He was so proud of this achievement that he asked to engrave this approximation on his tombstone. (This would be a difficult request for the mathematicians responsible for the current best-known approximation of π, as it is goes up to more than 100 trillion (10^{14}) decimal places.) A British amateur mathematician, William Shanks, was somewhat less fortunate. In the 19th century, he spent 15 years calculating π to 707 decimal places, but as was later discovered, made a mistake at the 528th place. Proving that π is irrational was a hard nut to crack and was established only in the 18th century.[10]

Another irrational number known to the Greeks was the *golden ratio*. It appears in many natural constructions. For example, as explained in Appendix 12, this is the ratio of the lengths of a diagonal and a side in a pentagram. Several astonishing applications of the golden ratio were found in music, nature, painting, and architecture, to name a few.

Aristotle and the foundations of logic One of the towering figures of Greek history was **Aristotle** (384–322 BCE), a teacher of Alexander the Great. For 20 years, he was a student at Plato's school, the *Academy*, and later founded his own school, the *Lyceum*. Aristotle wrote on a vast array of subjects, in particular, politics, ethics, and physics. He was also the first historian of philosophy, providing a systematic account of the works of previous philosophers. Some of his work can be seen as the origins of linguistics, the study of human languages.[11] His main contribution to mathematics was in providing the foundations of *logic*. In his work *Prior Analytics*, Aristotle argued that logical arguments should be constructed using *syllogisms*, inferences that allow one to conclude new statements from the already established ones. An example is the familiar deduction:

> Every Greek is a human.
> Every human is mortal.
> Therefore, every Greek is mortal.

Aristotle proposed 192 syllogisms in total, but the more known

> All men are mortal.
> Socrates is a man.
> Therefore, Socrates is mortal.

is not among them, since according to his logic, one could not make specific statements (like about Socrates). This syllogism seems to have been invented only in the High Middle Ages, probably by William of Ockham, a 13th-century

Oxford scholar (further discussed in Chapter 4) known for the 'Ockham's Razor' principle, who came up with 1,368 new syllogisms.[12]

According to Aristotle, the only way to attain valid scientific knowledge is by means of a logical argument based on syllogisms, starting with true statements, which are either *postulates* that hold for a particular science or *axioms* that always hold. This approach lies at the basis of scientific reasoning.

In the 3rd century BCE, Aristotle's approach to logic was taken further by the Stoics, who added *connectives* to logic, such as conjunction and negation. This allowed them, in contrast to Aristotle, to study complex propositions. The *modus ponens* rule (from two premises: A, and A implies B, infer B) is due to Stoics. Their approach to logic can be considered as the origin of what is now called *propositional logic*.

This view of logic was taken up 2,000 years later by Gottfried Wilhelm Leibniz. In turn, as I shall explain later, there is a direct (though lengthy) path that leads from Leibniz to computers. So, in a sense, we are all now profiting from Aristotle's insights.

Euclid's *Elements* The Pythagorean theorem is just one example of the many theorems that can be found in the *Elements*, a treatise by **Euclid** (c. 325–c. 265 BCE), a Greek mathematician from Alexandria, in which he compiled all mathematical knowledge known to the Greeks at that time. It consists of 13 volumes (called books) containing 465 theorems from geometry and number theory. Its English translation exceeds 500 pages.[13]

The *Elements* are unbearably dry and difficult to digest. So it is surprising that it is one of the most influential books in history. Also, almost nothing is known about its author. The story of how the *Elements* became available to us tells a lot about the convoluted ways of human civilization. The book was translated around 800 from Greek into Arabic in Baghdad by the Islamic mathematician al-Hajjaj ibn Matar, during the period called the Islamic Golden Age. Then, around 1120, an English monk called Adelard of Bath, a philosopher and a keen traveler through southern Europe, translated it from Arabic into Latin. Some 140 years later, Campanus of Novara, an Italian mathematician, used Adelard's second translation to produce an improved and annotated version. This version was first printed in Venice in 1482. The first translation of the *Elements* directly from Greek into Latin was carried out only in 1505 by an Italian Bartolomeo Zamberti. This version was, in turn, translated into English in 1570 by Sir Henry Billingsley, the later lord mayor of London. The original versions of the *Elements* still exist. For example, one is located in the Bodleian Library in Oxford.[14]

Throughout the ages the *Elements* have inspired and influenced famous artists, philosophers, scientists, and politicians. Euclid (together with Pythagoras, Socrates, Plato, Aristotle, and other well-known Greek figures) appears in the famous Renaissance painting, *The School of Athens* (1509–1510), by Raphael. Euclid is the person leaning forward at the bottom right of the painting.

Euclid in *The School of Athens*.[15]

The discovery of the *Elements* by the English philosopher Thomas Hobbes heavily influenced his thinking and convinced him that geometry was the key to the study of nature.[16] Dutch philosopher's Baruch Spinoza main work, *Ethics*, was written in a style inspired by Euclid's *Elements*.[17] Isaac Newton, an uncommonly solemn person, apparently laughed when an assistant asked him what benefit there might be in studying Euclid.[18] Bertrand Russell started to study the *Elements* at the tender age of 11, with his brother as his tutor. He commented on it: "This was one of the great events of my life, as dazzling as first love. I had not imagined that there was anything so delicious in the world."[19]

Given the length of the treatise, it is no wonder that Ptolemy, the ruler of

Egypt, asked Euclid whether there was an easier way to understand geometry than by studying the *Elements*. Euclid apparently replied: "There is no royal road to geometry."[20]

The first four books of the *Elements* discuss geometry and culminate in the construction of a regular pentagon that builds upon all geometry results established so far. Robin Hartshorne wrote about it in his guided reading of the *Elements*: "If there is such a thing as beauty in a mathematical proof, I believe that this proof of Euclid's for the construction of the regular pentagon sets the standard for a beautiful proof."[21] In Appendix 13, I provide a particularly simple construction found in the 19th century.

The Greeks held geometry in high esteem. Plato had the saying "Let nobody ignorant of geometry enter here" inscribed over the entrance to his school. In *The Republic*, he outlined a curriculum called the *quadrivium* that consisted of four studies: arithmetic, geometry, astronomy, and music. It was widely used in the Middle Ages.

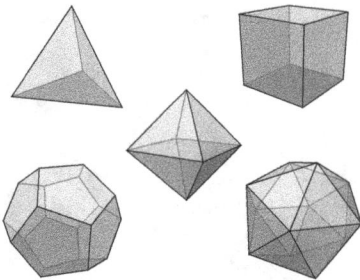

Platonic solids.[22]

In another book, *Timaeus*, Plato introduced five *Platonic solids*, which are regular and convex polyhedra (defined as three-dimensional objects enclosed by identical flat faces, the vertices of which lie on a sphere). In the 13th (and final) book of the *Elements*, it was proved that no other Platonic solids exist. A proof can be found in Appendix 21.

This striking result was described by Hermann Weyl, a prominent German mathematician of the 20th century, as "one of the most beautiful and singular discoveries in the whole history of mathematics".[23]

Euclid attempted to derive all his theorems about geometry from a short list of intuitive postulates and axioms, like the statements "all right angles are equal" or "the whole is greater than the part". It was noticed only at the end of the 19th century that he did not completely succeed since he overlooked some unstated assumptions.[24]

Euclid's 5th axiom One of Euclid's axioms, the 5th one, states (in an equivalent, modern form) that on a plane, given a straight line and a point not lying on it, one can draw through the point exactly one straight line parallel to the given line, as in the following drawing:

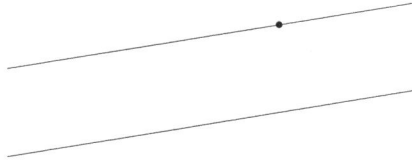

The 5th axiom of Euclidean geometry.

Over the centuries, mathematicians tried in vain to derive this axiom from the other Euclidean axioms. The matter, as we shall see later, was only clarified in the 19th century, which gave rise to *non-Euclidean geometry*, of which a variant was used in the 20th century in Einstein's general relativity theory.

Prime numbers The *Elements* also contains several important results about numbers, in particular about *prime numbers*. These are natural numbers greater than 1 that are divisible only by 1 and themselves. For instance 3, 7, and 23 are prime numbers while 12, 21, and 733,055,621 are not (the latter being the product of two prime numbers, 27,073 and 27,077). Euclid proved, in particular, that there are infinitely many prime numbers. Three proofs of this result can be found in Appendix 5.

Prime numbers have always kept the attention of mathematicians, but studying them seems like a useless occupation. However, in 1978, **Ron Rivest**, **Adi Shamir**, and **Leonard Adleman** proposed what is now called the RSA cryptosystem, which is widely used in secure data transmission, for instance, in credit card payments over the Internet. This system is based on the fact that multiplying two prime numbers is easy, but decomposing a product of them (like the abovementioned 733,055,621) into prime numbers can take a long time.

Prime numbers are deceptive in their simplicity. In fact, there exist simple to state problems about them that remain open. The most famous example is the *Goldbach conjecture* from 1742, which states that every even number larger than 2 is a sum of two prime numbers.

Three classical problems The Greeks are also responsible for the so-called *three classical problems* that ask for the following constructions using only a ruler and a compass:

- *Doubling the cube*: constructing a cube with the double volume of a given one,

- *Squaring the circle*: constructing a square of the area of a given circle,
- *Trisecting an angle*: dividing an angle into three equal parts.

The problems were formulated in the 5th century BCE. The first one is mentioned in Plato's *The Republic*. The second one occupied, among others, Leonardo da Vinci, who kept trying to solve it for ten years.[25] It was shown only in the 19th century that all three problems were unsolvable.[26] This has not deterred some people from working, sometimes for years, on these problems and submitting elaborate solutions to randomly chosen mathematicians. (This trend may have started with the philosopher Thomas Hobbes who kept claiming in 1665 that he solved the problem of squaring the circle in spite of the rebuttal by the mathematician John Wallis.) Those who deal with the trisection problem are called *trisectors*.[27]

Astronomic calculations   It is time to mention the most fateful mathematical mistake in history. Around 240 BCE, a Greek astronomer and mathematician, **Eratosthenes of Cyrene** (c. 276–c. 194 BCE), in a highly impressive achievement, computed almost exactly the circumference of the Earth. He did it by cleverly combining measurements of the shadows at noon at two distant locations with simple geometric reasoning. His argument is given in Appendix 6.

Unfortunately, some 140 years later, another Greek astronomer came to the wrong conclusion that the circumference was only about 29,000 kilometers and not 40,000 kilometers as stated by Eratosthenes. In the 2nd century, this mistake was incorporated into a famous world map of Claudius Ptolemy from Alexandria. For 13 centuries, this error persisted and eventually brought Christopher Columbus to the optimistic and erroneous conclusion that the westward voyage to India was feasible since it would be less than 5,000 kilometers. It was only after Magellan's crew circumnavigated the Earth in 1521–1523 that Eratosthenes's correct value was confirmed.[28]

Equally brilliant as Eratosthenes's calculations were those of **Hipparchus of Nicaea** (c. 190–c. 120 BCE), who some 90 years later, cleverly used information about the solar eclipse at two locations of a known distance: at one, it was total, while at another, only four-fifths of the Sun was obscured by the Moon. He combined this fact with a simple geometric reasoning to almost correctly compute the distance of the Moon from the Earth. Impressively, Hipparchus also determined the length of the year as $365 + \frac{1}{4} - \frac{1}{300}$ days, which differs from the correct length by only 6 minutes. He was also the first Greek scientist who used the Babylonian approach of dividing a circle into 360 degrees.

The approach of comparing the views of the same object from two different locations (or the same location at different times of the year) is called *parallax* and has become a basis for astronomic measurements.[29]

However, the computation of the Sun's distance to Earth was a bridge too far for the Greeks. Already in the century before Hipparchus, **Aristarchus of Samos** (c. 310–230 BCE) correctly used the parallax method to attempt this but got completely wrong results due to the limitations of available instruments. The first realistic estimate was provided only in the 17th century and involved simultaneous observations of Mars (it will become clear once we reach the account of the 17th century why Mars was observed) from two locations: Paris and in what is now French Guiana. The ultimate achievement of the parallax method took place in the 19th century, when it was used to compute the distance to a star.

Archimedes of Syracuse   Perhaps the greatest Greek mathematician was **Archimedes of Syracuse** (c. 287–c. 212 BCE), a contemporary of Eratosthenes (some of his letters to Eratosthenes survived). Besides being a mathematician, he was also an astronomer, a prominent engineer and an inventor, for example, of the *Archimedes screw*, which is still used today to pump water up from a lower-lying source. The perennial story of Archimedes has him, while taking a bath, shouting "Eureka!" upon discovering how to determine whether a crown was indeed made of pure gold.

In his work, Archimedes connected mathematics with the physical world, for example, by determining the center of gravity of a triangle. He also understood the mathematics behind the principle of a lever. The statement "Give me a place to stand on, and I will move the Earth" is attributed to him.

I already mentioned that Archimedes provided a good approximation of the value of π. He also established the formulas expressing the area of a circle, the surface area and the volume of a sphere, and the area under a parabola.[30] He achieved it by improving upon the *method of exhaustion*, which had been introduced more than 100 years earlier by **Eudoxus of Cnidus** (408–355 BCE). The idea of this method is to 'exhaust' the area of a figure by means of a successive approximation by polygons. Eighteen centuries later, this approach gave rise to *calculus*, a branch of mathematics concerned in particular with the computation in a more direct way of lengths, areas, volumes, and motion.

In the treatise titled *The Sand Reckoner*, Archimedes tried to calculate the number of grains of sand in the universe and devised for this purpose a counting system based on the number 10,000, called a *myriad*. Substantial new insight into Archimedes' work was provided thanks to a 13th-century prayer book that turned out to be written over two treatises of Archimedes. This manuscript has had a complicated history: on 16 July 1907, the front page of the *New York Times* reported its sensational discovery in Constantinople; unfortunately, the manuscript was lost after the First World War. Eventually, it made its way through France to the United States, where it was sold in 1998 at a Christie's auction for two million dollars. The anonymous buyer subsequently made it available to a museum in Baltimore, Maryland, US, where it has been extensively studied by historians, conservators, and mathematicians.[31]

E.T. Bell, a historian of mathematics, wrote in 1940: "Modern mathematics was born with Archimedes and died with him for two thousand years. It came to life again with Descartes and Newton."[32] It is appropriate to qualify this statement by mentioning a few important mathematicians who followed Archimedes.

Apollonius and Heron   During the secondary school, children learn about circles, *ellipses*, *parabolas*, and *hyperbolas*. These are the outcomes of an intersection of a cone and a plane and are therefore called *conic sections*.

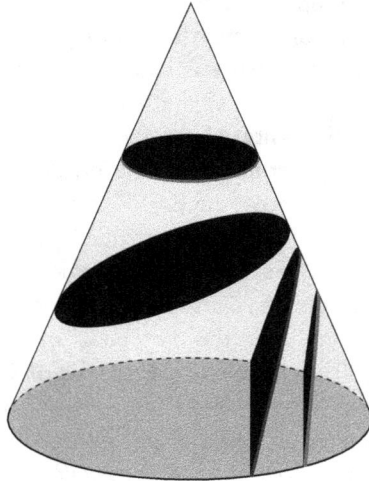

Conic sections.[33]

These conic sections were introduced around 350 BCE and studied extensively by **Apollonius of Perga** (c. 262–c. 190 BCE) in his book, *Conics*. It contains a wealth of material about them (389 theorems) and is considered to be one of the greatest works of Greek science. Both the ellipse and parabola will play a role some 1,800 years later, during the Scientific Revolution.

Apollonius also studied the following beautiful problem: given three circles, construct a fourth one that touches each of them. Here is an example of a solution for one case:

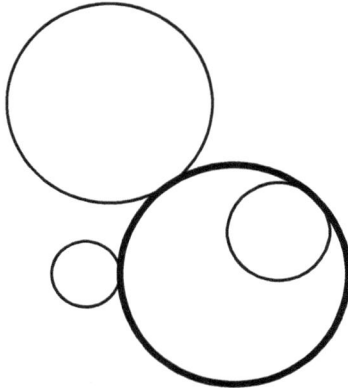

A solution to Apollonius' problem.

Apollonius' solution has since been lost. In the 17th century, an account of the problem was reconstructed, and this led to an intense study of it by many mathematicians, including Descartes. It has ten cases, depending on the relative location of the circles.

Moving forward three centuries, we encounter an impressive engineer, **Heron of Alexandria** (c. 10–c. 70 CE), the inventor of *Heron's engine*, the first steam engine, albeit used as a toy. His interest in engineering motivated him to formulate and solve, by means of geometry, the problem of constructing a straight tunnel through a mountain by workers digging it from both sides. In mathematics, he is remembered for two contributions: a method of computing a good approximation of the square root of a number and an elegant *Heron's formula* that expresses the area of a triangle in terms of its sides.[34] Also, he computed the volume of each of the Platonic solids.

Ptolemy, Diophantus and Hypatia The previously mentioned cartographer **Claudius Ptolemy** (c. 100–c. 170 CE) was a Greek who lived in Alexandria after it had been conquered by the Romans. He is mostly remembered for his fundamental works on astronomy and geography. I have already discussed his world map. In his treatise *Almagest* (meaning 'the greatest'), he described a detailed Earth-centered model of the universe, which, in particular included a catalog of 1,028 stars. This model of the universe persisted for 1,400 years. To produce it, he extensively used the works of Hipparchus and Apollonius and relied on trigonometry by producing what essentially corresponds to a table of values of the sine function. To do this, he came up with the following elegant theorem, now called *Ptolemy's theorem*. It states that in a quadrilateral inscribed in a circle, the product of the lengths of the diagonals equals the sum of the products of the lengths of the opposite sides. A novel proof of this theorem is given in Appendix 7.

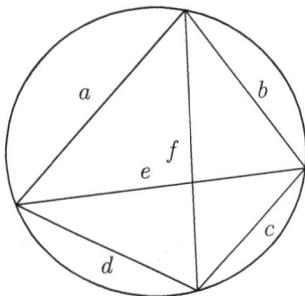

Ptolemy's theorem: $ac + bd = ef$.

Another mathematician worthy of mention is **Diophantus of Alexandria** (around 250 CE), sometimes called the 'father of algebra' because he was apparently the first mathematician to study equations. Little is known about him, and even his birth and death dates are uncertain, though a puzzle from the 6th century about his age yields '84' as the solution.

He published his findings in *Arithmetica*, a text to which I shall return in the account of the 20th century. *Arithmetica* consisted of 13 books. Six survived the 7th-century destruction of the library of Alexandria by the Muslims, and four more were discovered in 1968 in northeastern Iran, in an Arabic translation.

Diophantus introduced *unknowns* and was probably the first person who systematically made use of the equality sign (now written as '='). He studied, in particular, equations with integer coefficients for which one seeks integer solutions. Such equations are nowadays called *Diophantine equations*. He also investigated quadratic equations, though he did not provide a general method of solving them.

Diophantus's achievements were hampered by clumsy notation and his rejection of equations whose only solutions were negative numbers. Also, he did not provide any general method. Instead, each of the 189 problems he considered was solved by a different method.

Often mentioned in our account, Alexandria had been the major center of learning since the times of Euclid. It featured one of the largest libraries of the ancient world, which was part of the *Mouseion of Alexandria*, the first (in modern terminology) research institute. Among others, Euclid, Eratosthenes, and Archimedes were associated with it.

The beginning of the decline of this city is associated with the murder of the first known female mathematician, **Hypatia of Alexandria** (c. 355–415 CE). No information about her contributions survived, apart from the fact that she wrote commentaries on the works of Apollonius and Diophantus. Hypatia was brutally killed by members of a fanatical Christian sect who did not approve of her teachings of philosophy.[35] As Bertrand Russell wrote: "After this [her murder], Alexandria was no longer troubled by philosophers."[36] The library was destroyed two centuries later after a Muslim army captured the city.

Greek notation The Greeks represented numbers using letters of their alphabet. In particular, α stood for 1, β for 2, ξ for 60, and ϕ for 500. They did not know the concept of *zero*. Also, they did not use positional notation, so 61 could be represented both by $\alpha\xi$ and $\xi\alpha$. Fractions were written using the symbol $'$. So, $\frac{1}{3}$ was written as $\overset{\prime}{\gamma}$, and more complicated fractions were represented as appropriate sums, for instance, $\frac{7}{12}$ was represented as the sum $\frac{1}{3} + \frac{1}{4}$ yielding it, so $\overset{\prime\prime}{\gamma}\delta$. This system persisted for more than 1,000 years, until the fall of Constantinople in 1453.

Infinity One of the legacies of Greek mathematics is the concept of *infinity*, encountered by them in a number of ways. The simplest reference to it is in the result mentioned earlier that there are infinitely many prime numbers. But infinity is also implicitly present in many other ways. For instance, the Pythagorean theorem is a statement about *all* (so, infinitely many) right triangles. The fact that $\sqrt{2}$ is irrational means that it cannot be presented in a finite form of a fraction. Archimedes' use of polygons to approximate a circle also relies on the observation that each circle is the limit of an infinite sequence of polygons with an increasing number of sides.

Infinity was the subject of deep paradoxes attributed to **Zeno of Elea** (c. 490–430 BCE), the most famous of which is the *Achilles and tortoise* paradox, where Achilles tries to catch up with an escaping tortoise. The paradox states that Achilles can never succeed. Namely, he first has to reach the current location of the tortoise. By then, the tortoise has moved further along. Once Achilles reaches the new location, the tortoise moves further again, and so on. As explained in Chapter 4, behind this paradox lurks the idea of infinity—

more precisely, that an infinite sum can converge to a finite result. Aristotle discussed extensively the notion of infinity in order to refute Zeno's paradoxes.

The subject of infinity is, in fact one of the most crucial threads in the history of mathematics. It is at the heart of most practical questions, such as computing an area or a volume, or a representation of the outcomes of infinite calculations. Famous mathematicians of the Enlightenment and subsequent periods struggled with infinite sums and occasionally used wrong arguments when reasoning about them. It took mathematicians centuries to properly formalize their reasoning involving infinity and to capture various aspects of it in the correct way. Infinity is also the source of baffling questions, such as whether only one type of infinity exists (a question to which I shall return when discussing mathematics in the 19th century). In fact, as already indicated in the introduction, logicians and mathematicians today continue to debate the use of infinity in reasoning and the existence of infinite sets of a certain type.

Deductive method   Another legacy of the Greeks is the idea of a *deductive method*. According to it, results are supposed to be established from the original assumptions, by means of systematic reasoning in which all the steps consist of applying agreed upon rules to the original axioms and assumptions or to earlier established results. This was the essence of Euclid's approach in the *Elements*, in which he put forward his postulates and axioms. This sounds very much like Aristotle's presentation of logic. However, Greeks' proofs in number theory and geometry results were completely unrelated to the study of arguments in logic, as initiated by Aristotle. The reason was that the logics studied by Aristotle and later by Stoics were too poor to express even the simplest mathematical statements about numbers, let alone geometric figures.

These two threads—mathematics and logic—crossed only towards the end of the 19th century, when logic matured and extended the scope of its study.

Greek mathematics and today   Of the three main areas of mathematics —geometry, number theory, and logic— the Greeks contributed to, only the first one had obvious applications. It is noteworthy that in the end, with the advent of cryptography and computer science in the 20th century, all three turned out to be useful.

The symbolism and influence of Greek mathematics can be felt until today. Parabolas and ellipses are used in architecture, in particular, in the constructions of bridges and whispering galleries, while parabolic mirrors are used in astronomical telescopes. To 'square the circle' became a standard expression in many languages. The communist red star is just a filled pentagram, while the headquarters of the US Department of Defense is called the Pentagon because of the shape of the building.

Two Platonic solids, dodecahedron and icosahedron, are occasionally used as utility or art objects.

A lamp in the shape of a dodecahe-dron.[37]

A sculpture of Baruch Spinoza in Amsterdam, with an icosahedron.[38]

Paul Valéry, a 20th-century French poet and essayist, stated: "I never think of classical art without inevitably choosing the monument of Greek geometry as its best example."[39] In a book by Arthur Conan Doyle, Dr. Watson found that Sherlock Holmes' "conclusions were as infallible as so many propositions of Euclid".[40]

This concludes the account of Greek mathematics. In total, I discussed the contributions of a dozen mathematicians. Of course, several other Greek mathematicians contributed to the development of the discipline, but those mentioned here stand out in comparison to the others.

Timeline

Thales of Miletus (c. 625–545 BCE)
Pythagoras of Samos (c. 570–c. 500 BCE)
Zeno of Elea (c. 490–430 BCE)
Eudoxus of Cnidus (408–355 BCE)
Aristotle (384–322 BCE)
Euclid (c. 325–c. 265 BCE)
Aristarchus of Samos (c. 310–230 BCE)
Archimedes of Syracuse (c. 287–c. 212 BCE)
Eratosthenes of Cyrene (c. 276–c. 194 BCE)
Apollonius of Perga (c. 262–c. 190 BCE)
Hipparchus of Nicaea (c. 190–c. 120 BCE)
Heron of Alexandria (c. 10–c. 70 CE)
Claudius Ptolemy (c. 100–c. 170 CE)
Diophantus of Alexandria (around 250 CE)
Hypatia of Alexandria (c. 355–415 CE)

Notes

[1]Snellius is more known for the optical law of refraction named after him.

[2]E. Danson, *Weighing the World*, Oxford University Press, 2006.

[3]J. Bronowski, *The Ascent of Man*, British Broadcasting Corporation, p. 162, 1973.

[4]W. Dunham, *Mathematical Universe*, John Wiley & Sons, p. 99, 1994.

[5]E.S. Loomis, *The Pythagorean Proposition*, National Council of Teachers of Mathematics, 2nd edition, 1940. Available from http://www.eric.ed.gov/PDFS/ED037335.pdf.

[6]K. Basu, A new and rather long proof of the Pythagoras theorem, by way of a proposition on isosceles triangles, *College Mathematics Journal*, 47(5), pp. 356–360, November 2016.

[7]B. Russell, *History of Western Philosophy*, Unwin Books, p. 50, 1946.

[8]Formally, a *regular polygon* is a closed path connecting equally spaced points lying on a circle. A *regular pentagon* is a regular polygon with five sides.

[9]U.C. Merzbach and C.B. Boyer, *A History of Mathematics*, John Wiley & Sons, 3rd edition, p. 45, 2011.

[10]It was proved by J.H. Lambert in 1761.

[11]L.T.F. Gamut, *Introduction to Logic*, Volume 1, The University of Chicago Press, p. 10, 1991.

[12]Ockham's Razor is a methodological principle, usually explained as 'entities are not to be multiplied beyond necessities'. The number of syllogisms is cited after A. Gottlieb, *The Dream of Reason*, Allen Lane, p. 402, 2000.

[13]The most useful online version is due to David E. Joyce at http://aleph0.clarku.edu/~djoyce/java/elements/elements.html. A digitized version of the *Elements* is available at http://www.rarebookroom.org/Control/eucmsd/.

[14]J. Freely, *Alladin's Lamp*, Alfred A. Knopf, 2009, and U.C. Merzbach and C.B. Boyer, op. cit., p. 108.

[15]Courtesy: Public Domain, Wikipedia Commons.

[16]S.E. Stumpf, *Philosophy: History and Problems*, McGraw Hill, pp. 226–227, 1989.

[17] R. Goldstein, *Betraying Spinoza: The Renegade Jew Who Gave Us Modernity*, Schocken, p. 9, 2009.

[18] D.J. Boorstin, *The Discoverers*, Random House, Inc., First Vintage Books Edition, p. 410, 1985.

[19] B. Russell, *The Autobiography of Bertrand Russell*, Unwin Books, p. 30, 1975.

[20] U.C. Merzbach and C.B. Boyer, op. cit., p. 90.

[21] R. Hartshorne, *Geometry: Euclid and Beyond*, Springer, p. 50, 2000.

[22] Courtesy: Public Domain, Wikipedia Commons.

[23] H. Weyl, *Symmetry*, Princeton University Press, 1952.

[24] This was rectified by D. Hilbert in 1899; see, for example, J. Stillwell, *Mathematics and its History*, Springer, 3rd edition, p. 19, 2010.

[25] See W. Isaacson, *Leonardo da Vinci*, Simon & Schuster, p. 209, 2017.

[26] The impossibility of doubling the cube and of trisecting an angle was established by Pierre Wantzel in 1837 and the impossibility of squaring the circle by Ferdinand von Lindemann in 1882.

[27] See U. Dudley, What to do when the trisector comes, *The Mathematical Intelligencer* 5(1), p. 21, 1983.

[28] I. Asimov, *Asimov's New Guide to Science*, Penguin Books, p. 20, 1987.

[29] A.W. Hirshfeld, *Parallax: The Race to Measure the Cosmos*, W.H. Freeman and Company, p. 59, 2001.

[30] The first three formulas, taught at secondary school, are, respectively, πr^2, $4\pi r^2$, and $\frac{4}{3}\pi r^3$, where r is the radius of the circle or the sphere.

[31] There is a website devoted to this manuscript, http://www.archimedespalimpsest. org/. Its high-quality copy, together with an extensive commentary, was published as R. Netz, W. Noel, N. Wilson, and N. Tchernetska (eds.), *The Archimedes Palimpsest*, Cambridge University Press, 2011. For an account of its contents and history, see R. Netz and W. Noel, *The Archimedes Codex*, Weidenfeld & Nicholson, 2007.

[32] E.T. Bell, *The Development of Mathematics*, Dover Publications, p. 76, 1992 (originally published in 1940).

[33] A Tikz drawing adopted from https://tex.stackexchange.com/questions/457452/ draw-the-four-conic-sections.

[34] The formula is as follows. Let s be half of the perimeter of a triangle, i.e., $s = \frac{a+b+c}{2}$, where $a, b,$ and c are the sides of the triangle. Then, the area of the triangle equals $\sqrt{s(s-a)(s-b)(s-c)}$.

[35] For an account of the life and times of Hypatia, see M. Dzielska, *Hypatia of Alexandria*, Harvard University Press, Reprint edition, 1996.

[36] B. Russell, *History of Western Philosophy*, Unwin Books, p. 368, 1946.

[37] A photo taken by the author.

[38] A photo taken by the author.

[39] J.L. Heilbron, *Geometry Civilized: History, Culture, and Techniques*, Oxford University Press, p. 149, 2000.

[40] A. Conan Doyle, *A Study in Scarlet*, CreateSpace, p. 20, 2010, first published in 1887.

Chapter 3

Ancient Chinese and Indian Mathematics (From 5th Century BCE to 5th Century CE)

In parallel with the ancient Babylonian, Egyptian, and Greek civilizations, two other civilizations made significant contributions to the development of mathematics: the Chinese and the Indians.

Ancient Chinese In the period until the 5th century CE, Chinese scholars developed sophisticated techniques to solve specific concrete mathematical problems. This development was characterized by intermittent long periods of inactivity.

The Chinese were not interested in pursuing abstract mathematical notions for their own sake. Their emphasis was exclusively on empirical science. This resulted in a different approach to mathematics than the one developed by Greeks, who were also interested in purely theoretical questions (for instance, in proving that there are infinitely many prime numbers or determining the number of Platonic solids) and who developed a deductive method. Therefore, to understand the achievements of Ancient Chinese mathematics, one needs to focus on the specific problems they tackled.

Some contributions of Ancient Chinese scholars to mathematics may even predate Greek contributions. Unfortunately, none of the original ancient mathematical texts survived. This makes it difficult to determine the exact origins and timing of specific mathematical contributions.

One of the oldest surviving mathematical texts from China is *Zhou Bi Suan Jing* (The Mathematical Classic of the Zhou Gnomon), devoted to astronomy and mathematics. It is estimated that it was written in 500–200 BCE.[1] The text is written in the form of a dialogue between a historical figure from the 11th century BCE and a skilled mathematician. It contains an interesting drawing, which suggests an elegant proof of the Pythagorean theorem. It is discussed in Appendix 2.

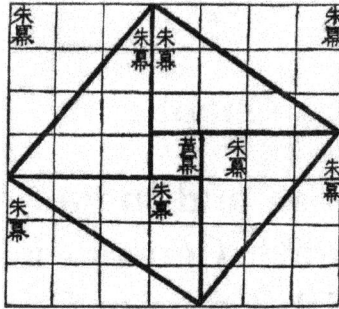

A drawing from *Zhou Bi Suan Jing.*[2]

The most influential text in the history of Chinese mathematics is an anonymous book, *Jiu Zhang Suan Shu* (The Nine Chapters on the Mathematical Art), written in 500 BCE–200 CE.[3] It is actually a compilation of works of several generations of Chinese mathematicians, possibly starting even from the 10th century BCE. The available version dates from the 3rd century and comes with a detailed commentary and a solution manual written by **Liu Hui** (c. 225–c. 295).

Liu Hui improved the text in a number of ways, in particular, by providing a better approximation of π. He also computed the volume of several solid figures, including a cone, pyramid, and tetrahedron (an arbitrary pyramid with a triangular basis), though he did not succeed in computing the volume of a sphere.

Chinese counting rod numerals.[4]

Several features make *Jiu Zhang Suan Shu* a remarkable text. It uses a decimal number system based on the so-called 'rod numerals' and additional symbols for the powers of 10. Thanks to their simple form, the numbers could easily be produced by using a bundle of bamboo sticks or scratching them in dirt. Given that the number system used by the Babylonians had base 60 and that the Egyptians did not have any signs for different digits, this strongly suggests that mathematics in China developed independently. Rod numerals have been used since the 8th century BCE, and for several centuries, it was the most advanced number system in the world.

Further, it is the first text in history in which negative numbers are used.

Additionally, rules for addition and subtraction in their presence were provided. Also, equations of a degree higher than three appeared. By comparison, these features were introduced in Europe only in the 15th century; see Chapter 5.

The text influenced Chinese mathematics in an analogous way as Euclid's *Elements* did for European mathematics, and for several centuries, it was used to train civil servants for the imperial bureaucracy. The book is organized as a series of 246 problems with solutions, arranged by topic into nine chapters.

The last chapter contains a proof of the Pythagorean theorem. It was, in particular, used to solve the following beautiful problem. Its attractive rendering, from the 13th century, is as follows, where *chi* is a length unit:

A 10-chi-high bamboo broke and its top touches the ground 3 chi from the base of the stem. At what height did it break?

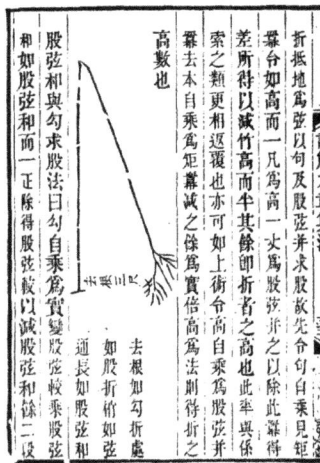

The broken bamboo problem.

A solution is given in Appendix 3. This problem is interesting because it reappeared in the works of Indian mathematicians from the 9th to the 12th century and eventually in some texts in Europe. This suggests a westward migration of a number of mathematical ideas from Ancient China.[5]

The sophistication of Chinese mathematics in the 4th century CE can be appreciated by considering the following mathematical problem from the text *Sun Zu Suan Jing* (Master Sun's Mathematical Manual):[6]

> There is an unknown number of objects. When counted in "threes"
> [i.e., divided by 3], the remainder is 2, when counted in "fives",
> the remainder is 3, and when counted in "sevens", the remainder
> is 2. How many objects are there?

The author of this book provided both an answer and an explanation of his method of solving the problem.[7] A solution, when generalized to an arbitrary number of assumptions of the form 'when counted in xs, the remainder is y', is called the *Chinese remainder theorem*. This result is highly relevant to the number theory. In particular, the RSA cryptosystem mentioned on page 17 relies on this theorem to ensure secure communication over the Internet. When discussing mathematics in the 20th century in Chapter 8, I shall mention another use of this theorem.

Another famous problem appeared in a text from the 5th century. It is called the 'hundred fowls problem' and is formulated as follows, where *qian* is a copper coin:

> If cockerels cost 5 qians each, hens cost 3 qians each, and 3 chick-
> ens cost 1 qian, and if 100 fowls are bought for 100 qians, how
> many cockerels, hens, and chickens are there?

The text offered three solutions to the problem, though no explanation was given on how they had been derived. Mathematically, this problem is interesting because it calls for solutions to linear equations in natural numbers. So, in the terminology introduced at the end of the previous chapter, the problem is concerned with solving Diophantine equations. Further, its formalization as two equations with three variables leads to the concept of *parameterized solutions*.[8] Interestingly, variations of this problem appeared a couple of centuries later in India, Egypt, and Medieval Europe.

In Chapter 2, I already mentioned a remarkable approximation $\frac{355}{113}$ of π found by a 5th-century Chinese mathematician **Zu Chongzhi** (429–500). He was also an astronomer, politician, and writer who made several other significant contributions to science. In particular, he found the formula for the volume of the sphere that eluded Liu Hui (and was discovered several centuries earlier by Archimedes). Also, he calculated the length of the solar year with phenomenal precision, comparable to the one provided by the Gregorian calendar introduced in the 16th century.

Of course, Chinese contributions to mathematics do not end in the 5th century. Mathematics continued to play an important role in Chinese society. In the 7th century, a collection called *Ten Computational Canons*, comprising ten Chinese mathematical texts, was compiled and subsequently used for imperial examinations. All three texts mentioned above formed part of this collection.

In the next chapter, I shall mention another significant contribution of a Chinese mathematician from the 13th century.

Ancient Indian Let us now move on to a short account of Ancient Indian mathematics. One of its striking characteristics was their fascination with numbers, both large and small.

The basis of Indian religion was the Vedas, extensive religious texts that were orally transmitted since 2000 BCE. From the beginning a counting system with base 10 was used. In one of the early texts, special names were given for powers 10 starting from ten all the way to a trillion (10^{12}). More extensive use of mathematics appeared in the appendices to the Vedas, called *Sulbasutras* (The Rules of the Cord).

One of the earliest Sulbasutras was written around 800 BCE by **Baudhayana**, about whom hardly anything is known. He may well have been the first mathematician in history whose name is recorded: recall that Thales lived almost 200 years later. The text contains geometric constructions leading to different approximations of π, the best one resulting in $\frac{900}{289}$, so approximately 3.114. It also states the Pythagorean theorem as:

> The rope [stretched along the length] of the diagonal of a rectangle makes an [area] which the vertical and horizontal sides make together.[9]

Note that this was almost 300 years before Pythagoras.

Historically, the most original is the following ingenious approximation of $\sqrt{2}$ using unit fractions (so—recall—fractions of the form $\frac{1}{n}$):

$$1 + \frac{1}{3} + \frac{1}{3 \cdot 4} - \frac{1}{3 \cdot 4 \cdot 34},$$

which is correct to five decimal places.

This attachment of mathematics to religion changed about 400 BCE, when works of mathematicians associated with one of the oldest Indian religions, Jainism, started to appear as independent texts. The interest in large numbers continued. In particular, Jain mathematicians devised a measure of time equal to the staggering number of $756 \cdot 10^{11} \cdot 8400000^{28}$ days.

They were also interested in the concept of infinity, by classifying numbers into three categories, which were, in turn, further subdivided into three classes:

- Enumerable: lowest, intermediate, and highest,
- Innumerable: nearly innumerable, truly innumerable, and innumerably innumerable, and

- Infinite: nearly infinite, truly infinite, and infinitely infinite.

This shows that Jain mathematicians discarded the idea that there is a single concept of infinity.[10] In Europe, such views were considered only in the 19th century in the visionary work of a German mathematician, Georg Cantor, who proposed a simple and far-reaching way of differentiating between various types of infinity; see Chapter 7.

One of the most original scientific developments in Ancient India was initiated by **Panini**, a scholar about whom very little is known. It is conjectured that he lived in the northwestern part of the Indian subcontinent between the 6th and 4th century BCE. Panini is considered the first scholar in linguistics. (His work predates the work of Aristotle.) Some 2,600 years ago, he published a book in which he came up with almost 4,000 structural rules that systematically described the structure of the grammar of "Classical" Sanskrit. In his work, Panini proposed about 1,700 basic building blocks that identified the basic components of the language, such as vowels, consonants, nouns, pronouns, and verbs, and grouped them into different categories. Subsequently, using appropriate suffixes and prefixes, he proceeded to construct compound words in the language.

This remarkable work is the first structural study of a language and has some natural connections with mathematics. Indeed, a derivation of a structure of a word or a sentence can be viewed as a validation, or a proof, obtained by means of the available syntactic rules. In Europe, Panini's work was discovered only in the 19th century. It influenced subsequent developments in linguistics.

Moving several centuries forward, we encounter **Aryabhata** (476–550), an astronomer and mathematician. The only text of his that survived is *Aryabhatiya*, a short treatise on astronomy and mathematics that integrated mathematical methods with astronomical explanations and summarized Hindu mathematics up to the 6th century. Calendar calculations contained in the text allowed one to infer that it was written in 499.

In *Aryabhatiya*, Aryabhata provided an incredibly accurate estimate of the circumference of the Earth, with a mistake of just 0.27%. It is not clear what method he used. He also found a strikingly good approximation of π, namely, 3.1416, which is correct to four decimals, though—again—it is not clear how he did it.

But the most important contribution in the mathematical part of the text is an algorithm motivated by a problem concerning two planets moving in opposite directions or the same direction. The task is to find the time of their meeting in the past and in the future, knowing their daily motions.[11] Such problems lead to similar considerations as those addressed by the Chinese remainder theorem mentioned earlier in this chapter. To solve them, Aryabhata provided an algorithm that was later called *Kuttaka* (the pulverizer). Recently, it was found that, just as the Chinese remainder theorem, *Kuttaka* is relevant for the RSA cryptosystem mentioned on page 17.

Aryabhata also devised a novel number system that was positional but had no digit zero. He used in it Sanskrit consonants for small numbers and vowels for successive powers of 10. In this system, numbers up to 10^{10} could be expressed using short phrases. For example, the number 57,753,336 was represented as the unpronounceable and meaningless word *cayagiyinusuchlr*.[12] This system did not catch on but possibly contributed to the acceptance of the idea of a positional number system.

The importance attached to Aryabhata's scientific contributions in India can be inferred from the fact that the first Indian satellite, launched in 1975, was named after him.

Two other Indian mathematicians played a crucial role in the early history of mathematics. Their contributions are discussed in the next chapter.

A knowledgeable reader may have noticed one important omission in this summary of Ancient Indian mathematics: the Hindu–Arabic numeral system that—as recently discovered—was invented in the 3rd or 4th century. However, given its far-reaching impact outside of India, it will be more convenient to discuss it in the next chapter.

Timeline

Baudhayana (around 800 BCE)
Panini (around 600–400 BCE)
Liu Hui (c. 225–c. 295)
Zu Chongzhi (429–500)
Aryabhata (476–550)

Notes

[1]G.G. Joseph, op. cit., p. 191.
[2]Courtesy: Public Domain, Wikipedia Commons.
[3]G.G. Joseph, op. cit., p. 191.
[4]Courtesy: Public Domain, Wikipedia Commons.
[5]G.G. Joseph, op. cit., p. 257.
[6]Quoted from G.G. Joseph, op. cit., p. 283.
[7]The smallest answer is 23. Indeed, 23 divided by 3 yields the remainder 2, 23 divided by 5 yields the remainder 3, and 23 divided by 7 yields the remainder 2.

[8]This problem can be formalized by means of the following two equations

$$5x + 3y + \tfrac{1}{3}z = 100,$$
$$x + y + z = 100,$$

where x, y, and z denote, respectively, the number of cockerels, hens, and chickens.

The solutions are of the form

$$x = 4t,$$
$$y = 25 - 7t,$$
$$z = 75 + 3t,$$

where t is an integer parameter.

All values, in particular y, have to be non-negative. It follows that t has to be 0, 1, 2, or 3. This leads to four solutions (x, y, z) in natural numbers: $(0, 25, 75)$, $(4, 18, 78)$, $(8, 11, 81)$, and $(12, 4, 84)$. The author missed the first one.

[9]G.G. Joseph, op. cit., p. 329.

[10]G.G. Joseph, op. cit., p. 350.

[11]The precise formulation is as follows:

> The two distances between two planets moving in opposite directions is divided by the sum of their daily motions. The two distances between two planets moving in the same direction is divided by the difference of their daily motions. The two results (in each case) will give the time of meeting of the two in the past and in the future.

(See W.E. Clarke, *The Aryabhatiya of Aryabhata: An Ancient Indian Work on Mathematics and Astronomy*, University of Chicago Press, p. 41, 1930. Republished by Kessinger Publishing, LLC in 2010.)

[12]G.G. Joseph, op. cit., p. 343.

Chapter 4

The Romans and the Middle Ages (From 1st Century BCE to the 15th Century)

To discuss mathematics in connection with the Romans, we need to step back a bit in time. The Roman civilization overlapped in time with the Greek civilization, and in more than one aspect, absorbed the latter's achievements and traditions. On its own, it hardly added anything to the development of mathematics. Its modest contributions to mathematics were indirect. Cicero, a Roman politician and famous orator, once provided the following assessment:[1]

> The Greeks held the geometer in the highest honour; accordingly, nothing made more brilliant progress among them than mathematics. But we have established as the limit of this art its usefulness in measuring and counting.

One of the Roman contributions was a new calendar, called the *Julian calendar*, introduced in 45 BCE by Julius Caesar. It was a solar calendar, and it replaced the previous Roman calendar, based on the moon phases. The new calendar stipulated that each year had 365 days and was divided into 12 months of different lengths. By some mistake, the leap years were initially introduced every three instead of four years, which caused a growing difference with the actual length of the solar year, which is close to $365\frac{1}{4}$ days. This error was finally addressed by Emperor Augustus, who ordered that the next three leap years be skipped. Eventually, the calendar was restored to its proper time in 8 CE.

The actual length of the solar year is about 365.2422 days, so the calendar year gradually shifted further and further away from the solar year. The growing discrepancy was taken care of only several centuries later. In 1582, Pope Gregory XIII decreed that the Julian calendar be replaced by the currently used *Gregorian calendar*. This calendar postulates that the years divisible by 100 are not leap years unless they are also divisible by 400. So, 1900 was not a leap year, while 2000 was. Some countries stuck to the Julian calendar for a couple of centuries longer. In particular, Russia adopted the Gregorian calendar only in 1918. To do this, 13 days had to be dropped, which explains why the October Revolution took place on 7 November 1917.[2]

Another invention attributed to Julius Caesar is the encryption of messages. To ensure the secrecy of military messages, he used a simple device that is now called the *Caesar cipher*. It replaces the original text by shifting each letter of the alphabet by a fixed number, say 3. So, for example, JULIUS CAESAR becomes MXOLXV FDHVDU (because J is replaced by M, U by X, etc.). If the recipient knows the fixed number shift, he can easily decrypt the message by reversing the encryption operation. To break this cipher, knowing that it was used, it suffices to try all 25 possibilities (assuming it was modeled upon the 26-letter English alphabet) for a shift. Still, this simple cipher was used as late as in the First World War by some troops of the Russian army. Needless to say, their messages were easily broken by Austrian and German cryptographers.[3]

The best-known contribution of the Romans to the history of mathematics is, of course, their notation for numerals in the form of the letters I, V, X, L, C, D, and M, which denote, respectively, 1, 5, 10, 50, 100, 500, and 1,000. (The Romans did not have a symbol for zero.) To parse numbers written in this notation in itself is not a problem. For instance, it is not so difficult to figure out that MDCCCLXXXVIII stands for 1,888. The problem is with manipulating them, for example, adding them up or, even worse, multiplying them is not for the faint-hearted. Moreover, this notation does not scale up: just try to write one million using Roman numerals.

Hindu–Arabic numeral system

This is why the Hindu–Arabic numeral system that we currently use is such a brilliant idea. Using this system, basic operations on numbers become dramatically simpler because of the positional notation used (recall that it means that a digit represents a different value depending on the position it stands). The crucial element of this system is the digit zero.

Its use goes back to the Bakhshali manuscript named after a village in what is now Pakistan, where it was found by a farmer in 1881. In the manuscript, ten digits were used, with a large dot representing zero. Until recently, it was thought that the manuscript was from the 8th century but 2017 carbon dating showed that it came from the 3rd or 4th century.[4] The manuscript is kept at the Bodleian Library in Oxford.

The first book that studied this numeral system was *Brahma Sphuta Sid-*

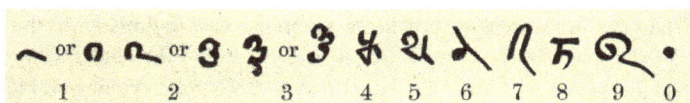

Ten digits from the Bakhshali manuscript.[5]

dhanta (Correctly Established Doctrine of Brahma), written by an accomplished Indian mathematician and astronomer **Brahmagupta** (598–670). He also made a number of novel contributions to algebra and geometry. In particular, he gave the first explicit formula for solving an arbitrary quadratic equation. His interest in mathematics for its own sake is revealed by an elegant formula that he established to determine the area of a quadrilateral inscribed in a circle (like the one considered in Ptolemy's theorem on page 22) in terms of its sides.[6]

Zero allows one to multiply a number by 10 in a trivial way, just by adding a '0' at the end. Continuing this way, we can multiply by 100, 1,000, etc. These multiplications by powers of 10 are implied by the positional notation, which allows us to represent numbers simply as a sequence of digits. For instance, to write two thousand and thirty-four, we do not need to write $2 \cdot 1000 + 3 \cdot 10 + 4$. We simply write the sequence of these four digits, 2034.

So, zero is much more than 'nothing': it is a device that, together with the positional system, revolutionized our way of using numbers. Another key advantage of using the Hindu–Arabic system is that the manipulations of numbers by means of the customary operations of addition, multiplication, subtraction, and division can be easily mastered.

The fact that this is all so obvious shows how deeply immersed we have become in this numeral system. It is fair to say that the Hindu–Arabic system substantially advanced civilization by providing it with the means of manipulating numbers, without which we could not properly function.

However, there is a small price to pay for the introduction of zero, which may perhaps explain why it appeared so late in history. After adding it to the realm of numbers, we also need to extend our rules for addition and multiplication. This is easy: we just postulate that for any number n, we have $n + 0 = 0 + n = n$, and $n \cdot 0 = 0 \cdot n = 0$. The subtraction is also straightforward, though it requires the acceptance of negative numbers: we postulate that $n - 0 = n$ and $0 - n = -n$. So far, so good.

Brahmagupta expressed such information for computing in the presence of zero and negative numbers by rules, such as the following ones, where 'fortune' represents a positive number, and 'debt' represents a negative number:

The product or quotient of two debts is one fortune.

The product of zero multiplied by a debt or fortune is zero.

But what about dividing by zero? Brahmagupta did not have a good suggestion for it, though he proposed that $\frac{0}{0} = 0$. In fact, the only meaningful solution is to disallow it.[7]

The use of ten digits is obviously motivated by our ten fingers: the Latin word for finger is 'digitum'. But it is useful to point out that the choice of ten is not crucial in the positional notation: it is easy to learn to add and multiply numbers in, say, binary notation or hexadecimal notation that uses 16 digits.

Al-Khwarizmi The Hindu–Arabic system was popularized in the Arab world by the Persian mathematician and astronomer *Al-Khwarizmi* (c. 780–c. 850), who worked in Baghdad during the Islamic Golden Age. (His full name is Muhammad ibn Musa al-Khwarizmi.)

He is worth mentioning for a couple of other reasons. His name in Latin, *Algoritmi*, is at the origin of the word 'algorithm', a procedure for describing a specific computation. Further, the word 'algebra' derives from the Arabic word 'al jabr', which comes from the title of Al-Khwarizmi's book, *Kitāb al-jabr wa al-muqābalah* (The Compendious Book on Calculation by Completion and Balancing), and refers to a specific operation on equations. In this book, he provided methods to solve linear equations and, independently of Brahmagupta, arbitrary quadratic equations.

Al-Biruni Another striking scholar from Central Asia was *Al-Biruni* (973–1048). He was familiar with the works of the Greeks, including those of Aristotle, Euclid, and Archimedes. Among others, he produced accurate sine and cosine tables and used this knowledge to compute in a brilliantly simple way the radius of the Earth, with a mistake of only 2%.[8] His argument is given in Appendix 8.

Arab scholars were interested in mathematics purely to advance the practical sciences they relied on, like astronomy or navigation. But they did not pursue it for its own sake. However, they also preserved Greek mathematics, combined it with Hindu mathematics, and eventually passed it back into Europe.

Bhaskara II One of the users of the Hindu–Arabic numeral system was an Indian mathematician *Bhaskara II* (c. 1114–1185). His work is viewed as the peak of mathematical knowledge in the 12th century.

Bhaskara II built upon the works of his Indian predecessor, Brahmagupta, which he extended in many ways. He wrote six works, the most known of which is *Lilavati* (The Beautiful), named after his daughter. The book dealt with various problems in algebra, arithmetic, and geometry. Interestingly, the broken bamboo problem mentioned in the previous chapter reappears in it. The book was translated into many languages, in particular, English, during the 19th century, which led to the popularization of various puzzles discussed in it.

Lilavati contains a study of certain Diophantine equations (equations with integer coefficients for which one seeks integer solutions; see page 22). In particular, Bhaskara II found a method that allowed him to find the least solution to the following innocently looking equation

$$61x^2 + 1 = y^2.$$

The solution is staggeringly large:

$$x = 226,153.980 \text{ and } y = 1,766,319,049,$$

for which both sides equal the astronomically large number

$$3,119,882,982,860,264,401.$$

While the used method was correct, its correctness was established only in 1929.[9]

In the 17th century, French mathematician Pierre de Fermat posed this problem as a challenge to a friend. In Europe, the problem was eventually solved some 100 years later by an Italian-French mathematician Joseph-Louis Lagrange.[10]

Leonardo Fibonacci The Hindu–Arabic numeral system was introduced in Europe only in the 13th century, by one of the most important mathematicians of the Middle Ages, **Leonardo Fibonacci** (c. 1170–c. 1250) from Pisa. He learned the system in Northern Africa from Arab merchants and scholars and advocated its use in his extensive book, *Liber Abbaci* (The Book of Calculation; the English translation has more than 600 pages), published in 1202.

The notation did not spread immediately. In particular, almost 100 years later, in 1299, it was banned in Florence. It took two centuries until the Hindu–Arabic system eventually superseded the one based on Roman numerals. The progress we have made since then can be best appreciated by reflecting for a moment on Samuel Pepys, a famous English diarist, who on 4 July 1662 reports his first lesson in learning the multiplication table. He was then 39; eight years earlier he had graduated from Cambridge University. Nowadays children learn the multiplication table during primary school.

However, Roman numerals did not completely die out. In some countries, like Poland and Russia, one still uses them to denote the months, so for example one writes 25 VII 2005 to denote 25 July 2005. Also, the names of kings, queens, and popes use Roman numerals, as in King Henry VIII or Pope John Paul II.

To some, analyzing consecutive numbers written in the Hindu–Arabic system became an obsession. In the 19th century, an Austrian mathematician Jakob Philipp Kulik, produced a table of all the numbers up to 100 million, together with their factors (so, in particular, a list of all prime numbers smaller than 100 million). It filled more than 4,000 pages. In turn, a Polish painter Robert Opałka painted practically every day, from 1965 until his death in 2011, consecutive numbers on canvases sized 196 by 135 centimeters, with each canvas containing several hundreds of numbers. At the end of his work, he exceeded the number 5.5 million. In 2010, Christie's sold three of his number paintings as a unit for $1.3 million.[11]

A page of the *Liber Abbaci*; on the right margin are the numbers
1, 2, 3, 5, 8, 13, 21, 34, 55, 89, 144, 233, 377
that form the beginning of the Fibonacci sequence.[12]

Motivated by a study of the growth of an idealized rabbit population, Fibonacci introduced in his book, *Liber Abbaci*, the famous *Fibonacci sequence*

$$1, 1, 2, 3, 5, 8, 13, 21, 34, 55, 89, 144, 233, \ldots$$

in which each element is the sum of the two preceding ones. This sequence is related to the golden ratio[13] and found many applications in mathematics. Also, it appears in nature, for example, in the arrangement of seeds on a sunflower. However, it took another 500 years before a formula was found to compute its elements in a direct way.[14] One keeps discovering new properties of Fibonacci numbers. There even exists a journal, *Fibonacci Quarterly*, founded more than 50 years ago.

The book also discussed the Chinese remainder theorem and included a version of the hundred fowls problem, both discussed in the previous chapter. This suggests an influence of, probably indirect, contacts with China. Incredibly, *Liber Abbaci* was translated into English only 800 years later.

Qin Jiushao Independently, an impressive text *Shu Shu Jiu Zhang* (Mathematical Treatise in Nine Sections), written by a mathematician **Qin Jiushao** (c. 1202–c. 1261), appeared in the 13th century in China. Qin was a man of many talents—a poet and an "expert at fencing, archery, riding, music, and architecture", but also described as "violent as a tiger, or a wolf, and as poisonous as a viper or a scorpion".[15]

His book contains 81 mathematical problems concerned with various practical issues, including astronomy, storage, and military matters. In particular, it contains a general formulation of the Chinese remainder theorem and a procedure on how to solve it. Qin Jiushao also rediscovered Heron's formula to compute the area of a triangle in terms of its sides, mentioned in the previous chapter. Further, remarkably, he explicitly used the digit zero which he denoted by a small circle.

One problem in his book is formulated as follows and can be best understood by consulting the corresponding figure; *li* is a distance unit.

> There is a round, walled town of unknown diameter with four gates. A tree lies 3 li north of the northern gate. If one walks 9 li eastward from the southern gate, the tree becomes just visible. Find the diameter of the town.

Depending on formalization, this problem leads to a third, fourth, or even fifth-degree equation, yielding the answer: 9 li. A solution to this problem is given in Appendix 17.

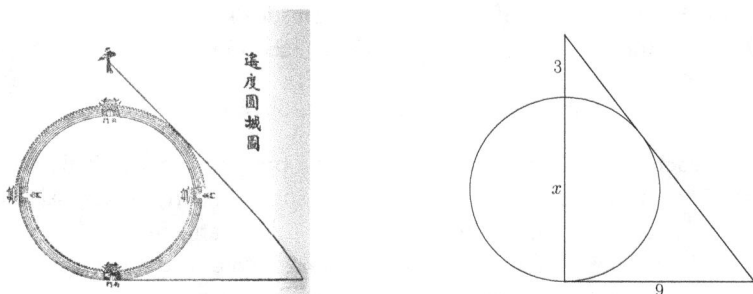

The round walled town problem and its representation.[16]

Though the text does not explain how the problem was solved, it does show that during the Middle Ages, Chinese mathematicians were well ahead of their European contemporaries in some areas. Qin was just one of the prominent 13th-century Chinese mathematicians. George Gheverghese Joseph, an expert on early non-European mathematics, wrote that in the subject of algebra, European mathematicians caught up with their Chinese counterparts only in the 18th century.[17] The colorful story of solving arbitrary third and fourth-degree equations by European mathematicians is discussed in the next chapter.

However, a remarkable decline, also in other sciences, took place in China, starting in the 14th century. Even some crucial mathematical texts were lost.[18] Many historians of science have since speculated why Europe, from the 16th century onwards, overtook China in the developments in science. Such a discussion is beyond the scope of this book. It is fair to say that Chinese mathematics recovered from this downturn only in the second half of the 20th century, after the Cultural Revolution.

Ramon Llull Returning to the Middle Ages in Europe, one should not omit the interesting figure of ***Ramon Llull*** (1232–1316), born on Mallorca, who during his long life was, among others, a philosopher, mystic, alchemist, missionary, university lecturer, an extremely prolific writer (he wrote more than 200 books), a monk (after an unhappy amorous affair), and a hermit. In his novel, the first one ever written in Catalan, the nuns in an abbey needed to elect a new abbess. They used a two-stage voting system, which one of them learned from a book of a certain Ramon Llull (perhaps a first 'cameo appearance' in the literature?). This novel and his two subsequent papers on the subject are now considered the beginnings of the voting theory, which is concerned with a mathematical study of voting methods. Llull's proposal was to compare candidates pairwise, an idea reinvented 500 years later by Marquis de Condorcet, to whom I shall return.

Llull's lifelong mission was to convert Arabs to Christianity. This required the mastering of Arabic, so (how else?) he bought a Muslim slave from whom he learned Arabic fluently. In arguing the virtues of the Catholic religion, he came up with a set of basic attributes of God or general principles (18, 16, or 9, depending on the version) and wrote short essays on each pair.

A diagram with 18 general principles (Goodness, Greatness, Eternity, Power, Wisdom, Will, Virtue, Truth, Glory, Difference, Concordance, Contrariety, Beginning, Middle, End, Majority, Equality, and Minority) from the 1517 edition of Ramon Llull's *Ars Magna* (Great Art).[19]

Llull also applied this way of combining features to other aspects of human knowledge, for instance, to medicine or psychology. For example, he compared the seven vices and seven virtues by the appropriate number of essays. This sounds eccentric, but it is still an early example of the use of combinatorics, an area of mathematics concerned with counting and generating various combinations of elements of a finite set, for instance, the set of all permutations.[20]

Nicolas Oresme Mathematicians starting from the Middle Ages began to experiment with infinite sums. Eventually, a couple of centuries later, this led to the invention of the calculus. Let me illustrate the matter by returning to Zeno's paradox from Chapter 2 in order to compute the distance Achilles has to cover to reach the turtle. Assume that Achilles is twice as fast as the turtle and that the initial distance is 1 (say, meter). Achilles needs to cover 1 first. During this time, the turtle moves $\frac{1}{2}$ away. When Achilles covers $\frac{1}{2}$, the turtle moves half of this distance, so $\frac{1}{4}$ away. And so on. This yields the infinite sum

$$1 + \frac{1}{2} + \frac{1}{4} + \frac{1}{8} + \dots$$

that converges to 2. This fact was already known to Archimedes.

However, **Nicolas Oresme** (c. 1323–1382), a French philosopher, astronomer, astrologer, and mathematician, came up in around 1350 with another infinite sum

$$1 + \frac{1}{2} + \frac{1}{3} + \frac{1}{4} + \dots$$

and showed, using a simple argument, presented in Appendix 9, that it diverges. The importance of this observation is that the successive terms become arbitrarily small, yet their sum—in contrast to the sum representing Zeno's paradox—is infinite. The successive partial sums of this infinite sum are called *harmonic numbers* because of their relation with the wavelengths of the vibrating strings.

Oresme also came up with the idea of representing the change of velocity in time using coordinates on a plane, an idea brought much farther three centuries later by Descartes. He was also the first mathematician who started using the plus sign $(+)$.

The progression in adopting the currently used mathematical symbols was agonizingly slow. The minus symbol (-) started to be used around 1489, and the equality sign $(=)$ in 1557, while the multiplication sign (written as $\times$) dates from 1618.[21]

Summary The period discussed in this chapter covers some 1,400 years, from Julius Caesar to Nicolas Oresme, so it is longer than the one spanned by Greek mathematics. If we ignore Greek contributions to mathematics, which ended in the 5th century, it is sobering to note how little—notwithstanding the contributions of Chinese and Indian mathematicians—was achieved in mathematics in these 14 centuries.

The Middle Ages cover the period from the 5th to the 15th century. Ivor Grattan-Guinness, a historian of mathematics, titled the corresponding chapter in his history of mathematics, 'A quiet millennium'.[22] For a couple of centuries after the fall of the Western Roman Empire in 476 CE, developments in mathematics took place thanks to Chinese, Indian, and Arab civilizations. In the period 600–1050, Europe was so much behind in comparison with the Arab civilization that a 10th-century Arab geographer reported that Europeans have "large bodies, gross natures, harsh manners, and dull intellects [. . .] those who live farthest north are particularly stupid, gross and brutish".[23] The sad thing is that during this period, one of the main uses of mathematics in Europe was in astrology. It was only during the Renaissance that mathematics rose above the level set by the Greeks.

Things began to change in the High Middle Ages when towns started to grow, which among others, led to the founding of Universities. The first one was established in Bologna in 1088. Oxford followed in 1096, Salamanca in 1134, Paris in 1160, and Cambridge in 1209.

The rise of Universities had a positive influence on the developments in logic. Among a dozen of scholars who contributed to the subject from the 11th to the 15th century, let me just mention a Franciscan friar **William of Ockham** (c. 1287–1347), who is already discussed in Chapter 2. After completing his studies of theology at the University of Oxford, he wrote in Latin an almost thousand-page-long treatise on logic, *Summa Logicae* (Sum of Logic). In it, following Aristotle, he studied syllogisms, but he also went much further by investigating the logical relationship between propositions. This led him to a more general concept of *inference*, which provided the early foundations of the modern proof theory. He also substantially extended Aristotle's study of so-called modal syllogisms.

This was one of the earliest contributions to the *modal logic* concerned with reasoning in the presence of modalities, such as 'possibly' and 'necessarily'. In the 20th century, modal logic was vastly extended to deal with reasoning about temporal relations, permissions, obligations, beliefs, and knowledge. It also led to unexpected applications in computer science, where it was used to formally verify computer programs.

Timeline

Brahmagupta (598–670)
Al-Khwarizmi (c. 780–c. 850)
Al-Biruni (973–1048)
Omar Khayyam (1048–1131) (mentioned in the next chapter)
Bhaskara II (c. 1114–1185)
Leonardo Fibonacci (c. 1170–c. 1250)
Qin Jiushao (c. 1202–c. 1261)
Ramon Llull (1232–1316)
William of Ockham (c. 1287–1347)
Nicolas Oresme (c. 1323–1382)

Notes

[1]V.J. Katz, *A History of Mathematics, an Introduction*, Addison-Wesley, 3rd edition, p. 157, 2009. Quoted after J. Sesiano, *Books IV to VII of Diophantus' Arithmetica in the Arabic Translation Attributed to Qusta ibn Luqa*, Springer-Verlag, 1982.

[2]D.E. Duncan, *The Calendar*, Fourth Estate Limited, 1998. The Gregorian calendar is still not completely accurate but the discrepancy of one day will show only in 4915. Namely, in each period of 400 years, starting from 1600, there are 97 leap years. This translates to an average addition of $\frac{97}{400} = 0.2425$ day per year. The difference of 0.0003 day with the solar year will eventually grow to one day after 3,333 years.

[3]D. Kahn, *The Codebreakers: The Comprehensive History of Secret Communication from Ancient Times to the Internet*, Simon & Schuster, pp. 631–632, 1996.

[4]See Carbon dating reveals earliest origins of zero symbol, BBC News, 14 September 2017, http://www.bbc.com/news/uk-england-oxfordshire-41265057.

[5]Courtesy: Public Domain, Wikipedia Commons.

[6]The formula generalizes Heron's formula for expressing the area of a triangle in terms of its sides, given in Chapter 2. Let s be half of the perimeter of a quadrilateral inscribed in a circle, i.e., $s = \frac{a+b+c+d}{2}$, where $a, b, c,$ and d are the sides of the quadrilateral. Then its area equals $\sqrt{(s-a)(s-b)(s-c)(s-d)}$.

[7]This is a consequence of the fact that division is defined as the converse of multiplication; that is that $\frac{a}{b} = c$ means that $a = b \cdot c$.

Indeed, note first that division is uniquely defined, for if $\frac{a}{b} = c_1$ and $\frac{a}{b} = c_2$, then $0 = \frac{a}{b} - \frac{a}{b} = c_1 - c_2$, i.e., $c_1 = c_2$. Next, suppose that for some a and c, we have $\frac{a}{0} = c$. Then $a = 0 \cdot c$, so $a = 0$. This means that $\frac{a}{0}$ can be meaningfully defined only for $a = 0$. However, $\frac{0}{0}$ is not uniquely defined, since from $0 = 0 \cdot 0$ and $0 = 1 \cdot 0$, we conclude that both $\frac{0}{0} = 0$ and $\frac{0}{0} = 1$.

Allowing division by zero also leads to fallacious proofs, like this one. Suppose $a = b = 1$. Then $a^2 - b^2 = a - b$, so $(a+b)(a-b) = a - b$ and hence $a + b = 1$, i.e., $2 = 1$.

For a historical account of this problem, see S. Saitoh, *History of division by zero and division by zero calculus*, International Journal of Division by Zero Calculus, 1(1), pp. 1–38, 2021.

[8]A. Sparavigna, The science of Al-Biruni, *International Journal of Sciences*, 2(12), pp. 52–60, 2013.

[9]V.J. Katz, op. cit., p. 249.

[10]G.G. Joseph, op. cit., p. 392.

[11]W. Grimes, Roman Opalka, an artist of numbers, is dead at 79, *New York Times*, 9 August 2011, http://www.nytimes.com/2011/08/10/arts/design/roman-opalka-conceptual-artist-with-numerical-focus-is-dead-at-79.html.

[12]Courtesy: Public Domain, Wikipedia Commons.

[13]As proved more than 300 years later by Johann Kepler, the ratio of consecutive Fibonacci numbers converges to the golden ratio.

[14]It was done in the 18th century by a French-English mathematician, Abraham de Moivre.

[15]J.J. O'Connor and E.F. Robertson, Qin Jiushao, https://mathshistory.st-andrews.ac.uk/Biographies/Qin_Jiushao, 2003.

[16]Courtesy for the first picture: Public Domain, Wikipedia Commons.

[17]G.G. Joseph, op. cit., p. 196.

[18]G.G. Joseph, op. cit., p. 292.

[19]Courtesy: Public Domain, Wikipedia Commons.

[20]J. Gray, "Let us calculate!": Leibniz, Llull, and the computational imagination, *The Public Domain Review*, 10 November 2016, http://publicdomainreview.org/2016/11/10/

let-us-calculate-leibniz-llull-and-computational-imagination/ and D.E. Knuth, *Art of Computer Programming, Volume 4, Fascicle 4: Generating All Trees—History of Combinatorial Generation*, Addison-Wesley, pp. 56–59, 2016.

[21]See the table of mathematical symbols at https://en.wikipedia.org/wiki/Table_of_mathematical_symbols_by_introduction_date.

[22]I. Grattan-Guinness, *The Rainbow of Mathematics*, W.W. Norton & Company, 2000.

[23]R.E. Lerner, S. Meacham, E. McNall Burns, *Western Civilizations*, W.W. Norton & Company, 13th edition, p. 261, 1998.

Chapter 5

The Early Modern World
(From the 15th to 17th Century)

The Renaissance started in Italy in the 14th century, though the term was coined only by a 16th-century Italian art historian, Giorgio Vasari. In this period, classical Greek and Roman manuscripts were rediscovered, and a revival in art, architecture, and literature took place. Indirectly, it also contributed to progress in mathematics.

Piero della Francesca (c. 1415–1492) was an Italian painter of the early Renaissance, known for probably the first application of perspective to paintings, an invention devised around 1415 by the Florentine architect Filippo Brunelleschi. Less known is that Piero della Francesca was also a mathematician who came up with a clever formula for the volume of a tetrahedron expressed in terms of its six edges. It generalizes the previously mentioned Heron's formula concerning the area of a triangle to three dimensions.[1]

An important book from this period is *Summa de Arithmetica, Geometria, Proportioni et Proportionalita* (Summary of Arithmetic, Geometry, Proportions, and Proportionality), written by an Italian friar *Luca Pacioli* (c. 1447–1517), who became the first holder of the mathematics chair at the University of Perugia. In the book, written in Renaissance Italian, he summarized all the mathematical knowledge available at that time and also introduced the double-entry book-keeping system, which is at the origin of accounting. This part of the book was in such demand that it was extracted and published separately.

In another book *De Divina Proportione* (Divine Proportion), illustrated by Leonardo da Vinci, the golden ratio (as the Italian title suggests) and the use of perspective in paintings are discussed extensively among others.

The knowledge of perspective was introduced in Northern Europe by *Albrecht Dürer* (1471–1528), who learned it during his stay in Italy. Dürer is considered to be the greatest German Renaissance artist. He was keenly interested in mathematics. In particular, he read Euclid's *Elements* and Pacioli's account of perspective, and he wrote the first book on mathematics in German. His most famous work is the mysterious engraving *Melencolia I*. In the background of the image below (left) one can see a polyhedron (a three-dimensional object with flat faces and straight edges), and in its upper right-hand corner, a 4 by 4 square with the year (1514) of the engraving, put in the middle of the bottom row (right). The square is filled with consecutive numbers $1, \ldots, 16$; the numbers lying in each row add up to the same number, 34, and the same is the case for each column and both diagonals. Such squares are called *magic*. Other striking properties of this square were noticed; for example, the sum of all numbers in each of the four quadrants also equals 34.

Dürer's *Melencolia I* and its magic square.[2]

The story of solving third and fourth-degree equations

One of the important concerns of mathematicians in the early 16th century was the problem of solving equations. The Babylonians already knew how to solve specific *quadratic equations*. As mentioned in the previous chapter, the solutions for an arbitrary case (now taught in secondary school) were first published by Brahmagupta, in the 7th century, and by Al-Khwarizmi, in the 9th century.

The story of finding the methods of solving *third* and *fourth-degree* equations involves Italian mathematicians and features probably the first dispute about scientific priorities. By *solving*, we mean the task of representing the solutions using the customary four arithmetic operations and the roots of arbitrary degrees (for instance, $\sqrt[3]{2}$ is a solution of the equation $x^3 = 2$).

At the end of his *Summa*, Pacioli stated that a solution of a third-degree equation is as impossible as the problem of squaring the circle. This remark stimulated **Scipione dal Ferro** (1465–1526), who eventually found a way to solve certain equations of the third degree. He kept his method secret, though he passed it on to a student of his. But the information about his discovery spread, and after his death, **Niccolo Tartaglia** (1499/1500–1557) found a method of solving arbitrary equations of the third degree. He also played an important role in the history of mathematics for another reason: his translation of Euclid's *Elements* into Italian was later studied by Galileo.

Tartaglia exploited his knowledge in a public contest—a regular event of the times—against a student of dal Ferro, who was unable to solve any problem submitted by Tartaglia. The contest caught the attention of **Gerolamo Cardano** (1501–1576), a colorful figure who was a physician and a mathematician with a keen interest in astrology, and who earned money by casting horoscopes and by gambling. Tartaglia eventually succumbed to the charms of Cardano and confessed his method to Cardano, though he made him swear to keep it secret.

But sometime later, Cardano learned that Tartaglia was preceded in his discovery by dal Ferro, so he decided to discuss the method (acknowledging Tartaglia) in his highly influential book *Ars Magna* (Great Art) that played an important role in the history of algebra. In the book Cardano also published a method of solving fourth-degree equations, credited to his former secretary, **Lodovico Ferrari** (1522–1565). Ferrari's method crucially depended on Tartaglia's method as it proceeded by a clever reduction of the problem to that of solving equations of the third degree.

Tartaglia was incensed that Cardano did not keep the secret, as he had planned to publish his findings in a book of his own. The quarrel that ensued was continued on behalf of Cardano by Ferrari and led to a lively and vitriolic correspondence between the two and a public contest that was won by Ferrari.[3]

A post scriptum to this story is that unknowingly to all involved, almost 500 years earlier, **Omar Khayyam** (1048–1131), a revered Persian poet, mathematician and astronomer, had already discovered how to solve almost all third-degree equations by geometric means that he had mastered by studying the *Conics* book of Apollonius.[4]

Once the methods of solving third and fourth-degree equations were discovered, it was natural to move on to the fifth-degree equations. In spite of the concerted effort of many prominent mathematicians, the problem of solving them remained open for almost 400 years. It is one of the most dramatic stories in the history of mathematics, to which I shall return in Chapter 7.

Negative numbers   As mentioned earlier, negative numbers were introduced by Chinese mathematicians. Brahmagupta used them in the 7th century. In Europe, they became known through Arab texts, but their acceptance was slow. Mathematicians of the Renaissance widely distrusted negative numbers and dealt with them in a roundabout way. Some of them spoke of negative numbers as absurd numbers.

As late as the 17th century, Blaise Pascal considered the subtraction of 4 from 0 as nonsense.[5] Consequently, Tartaglia and others, instead of considering an equation such as $x^3 - 15x - 4 = 0$, would rather analyze $x^3 = 15x + 4$. This led, of course, to several cases, depending on whether (using modern terminology) the used coefficients were positive or negative. Nowadays, we have no mental problem with negative numbers (or the minus sign) and need to consider just one general equation of the third degree.

Discovery of complex numbers   This digression into negative numbers should allow us to appreciate the boldness of the discovery of complex numbers, an event as momentous in the history of mathematics as the discovery of irrational or negative numbers. It was also the first example in the history of mathematics when an entity not directly related to the observed world was introduced.

The subject of complex numbers is also the first topic in the history of mathematics that goes beyond high-school arithmetic. The technical details can be skipped.

One complication concerned solving the third-degree equations using the method published by Cardano was that it led to some difficult to understand outcomes. In particular, it is straightforward to check that 4 is a solution of the equation $x^3 = 15x + 4$, but Tartaglia's method led instead to a strange outcome, namely

$$\sqrt[3]{2 + \sqrt{-121}} + \sqrt[3]{2 - \sqrt{-121}},$$

which involves taking the square root of the negative number -121.

Raphael Bombelli (1526–1572) found a way to reduce this strange expression to $2 + \sqrt{-1} + 2 - \sqrt{-1}$, which still involves suspicious roots, but—

miraculously—yields, as desired, 4. He did it by ignoring the fact that the roots of negative numbers do not exist and treating them instead as 'legal' numbers.

In the last year of his life, he published an influential book, *L'Algebra*, in which he proposed multiplication rules concerning positive and negative numbers. He listed expected entries like 'piu via meno fa meno' (positive times negative makes negative), but also—crucially—the multiplication rules concerning $\sqrt{-1}$, which we now denote by i, and that he called 'piu di meno' (more than less), and $-i$, that he called 'meno di meno' (less than less), which showed that he really understood the issues involved.[6]

Once i is 'admitted', the square roots of other negative numbers can be readily expressed using it. For example, the earlier mentioned $\sqrt{-121}$ is just $\sqrt{121}i$. Nowadays, numbers that are expressed using i, such as $3.5 + 4i$, are called *complex numbers*. Thanks to Bombelli's multiplication rules, complex numbers can be added, subtracted, and multiplied, just like the 'usual' real numbers. Also, the division of two complex numbers can be defined so that it becomes the converse of multiplication. Now it is only a mental problem to accept them because from an algebraic point of view, there is nothing strange about them. A minimal introduction to complex numbers is given in Appendix 10.

When complex numbers were discovered, they seemed to be a useless generalization of real numbers only needed to represent solutions of some equations. Not surprisingly, mathematicians originally had difficulties accepting them. More than 200 years later, a natural representation of them was found that makes them a bit less mysterious. In this representation, complex numbers are viewed as points on a plane, just like how real numbers can be viewed as points on a real line. For example, the complex number $3.5 + 4i$ corresponds to the point with the coordinates 3.5 and 4.[7] This calls for a broader understanding of the concept of a number, namely, that it can be either a one or a two-dimensional entity. And since the horizontal line passing through the 0 point of the vertical line is part of a plane, the real numbers naturally become a special case of the complex numbers, just like how the rational numbers are a special case of the real numbers.

From the 19th century, complex numbers began to be used in physics and electrical engineering, when dealing with entities that can be represented as a pair of related quantities.[8] They also have played a crucial role in establishing deep results in number theory and geometry that will be mentioned later, where some results concerning natural numbers or reals are obtained by a 'detour' through complex numbers. This is analogous to Bombelli's discovery that 4 is a solution of the equation $x^3 = 15x + 4$ by a detour through complex numbers.

Rise of modern algebraic notation　　　Mathematicians of the 16th century learned how to deal with equations of the third and fourth degree, but they used a truly clumsy notation. For example, Cardano would write '1.cubus.p.3.pos.p.2.aeq.0' for the equation $x^3 + 3x + 2 = 0$ and Bombelli would write 'Rc.2pRq.om121' for $\sqrt[3]{2 + \sqrt{-11}}$, which not only made the reading strenuous but complicated simple algebraic manipulations.[9]

The slow but steady progress in mathematical notation can be appreciated through these revealing examples:[10]

Regiomontanus, 1464:

　　3 Census et 6 demptis 5 rebus aequatur zero.

Pacioli, 1494:

　　3 Census p 6 de 5 rebus ae zero.

Stevin, 1585:

　　$3② - 5① + 6◯ = 0$.

Viète, 1591:

　　3 in A quad -5 in A plano $+$ 6 aequatur 0.

Descartes, 1637:

　　$3x^2 - 5x + 6 = 0$.

From this sequence, one can see that particularly significant improvements were achieved by Stevin and Viète independently.

Simon Stevin (1548–1620) was a Flemish-Dutch mathematician best known for having introduced the decimal point. The idea of decimal expansion led him to state something we now consider as obvious; namely, that all quantities such as natural numbers, square roots, rational numbers, irrational numbers, and negative numbers should be all viewed as a single 'species', now called real numbers.

Stevin was fully aware of the importance of his idea of the decimal point. He published his proposal in a short booklet titled *De Thiende* (The Tenth) that proudly announced on its cover page that it was "for astrologers, surveyors and measurers of tapestries".[11]

This reference to possible applications reflects Stevin's broad interests. He held no university position and earned money through various projects that reflected contemporary concerns, like improvements to windmills and sluices. Always a creative man, he came up with a number of interesting inventions, like a carriage with sails that achieved on a windy day at a beach close to the Hague the impressive speed of 40 kilometers per hour. Also, he was the first who explained that the tides rose due to the attraction of the Moon.

D E

THIENDE

Leerende door onghehoorde lichticheyt
allen rekeningen onder den Menfchen
noodich vallende , afveerdighen door
heele ghetalen fonder ghebrokenen.

Befchreven door S I M O N   S T E V I N
van Bruggbe .

T O T L E Y D E N,
By Chriftoffel Plantijn
M. D. LXXXV

Title page of *De Thiende* (1585).[12]

François Viète (1540–1603) was a French lawyer. He was the first to systematically use the + and − signs and to denote constants and unknowns by letters, and he is considered the 'father of modern algebraic notation'. Viète was a member of the king's council attached to the courts of Henri III and Henri IV. In that capacity, he was impressively successful in deciphering enemy messages. He treated mathematics as a hobby, yet made a number of important contributions to algebra and geometry. His book *In Artem Analyticien Isagoge* (Introduction to the Analytic Arts) is considered the starting point of modern algebra.

Gerardus Mercator   An unrelated contribution from this period involved a novel application of geometry. ***Gerardus Mercator*** (1512–1594), a German brought up in Flanders (now Belgium), was a cartographer and a pupil of the previously mentioned Gemma Frisius. He traveled little, which did not deter him from producing in 1569 a famous world map using a new method, now called the *Mercator projection*.

The method was based on a geometric projection of the Earth on a cylinder

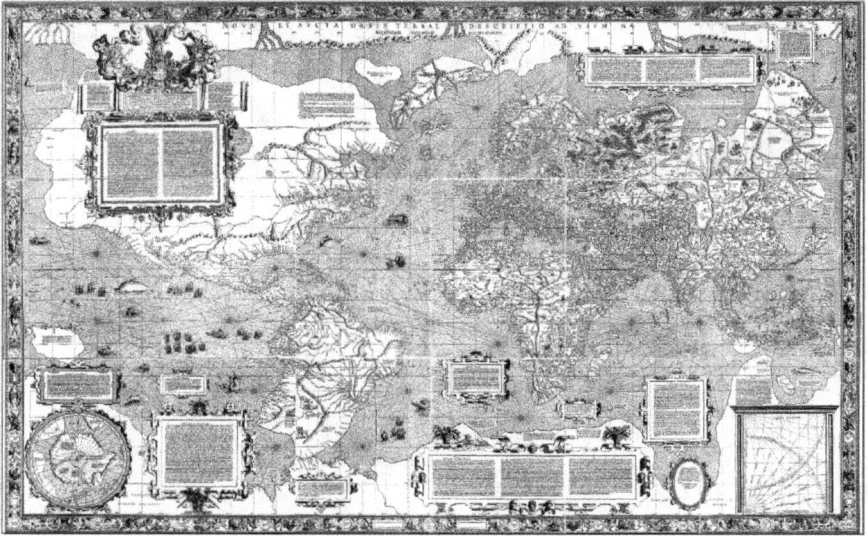

The 1569 Mercator map of the world.[13]

wrapped around it at the equator. After unrolling the cylinder, one gets a map. Mercator projection became a standard method of producing maps and is still used today.

Nicolaus Copernicus   The period being discussed covers the *Scientific Revolution*, a term used to denote the emergence of modern science. Its beginning is marked by a proposal of a new model of the universe by a Polish amateur astronomer **Nicolaus Copernicus** (in Polish **Mikołaj Kopernik**) (1473–1543). He published his findings in his book *De revolutionibus orbium coelestium* (On the Revolutions of the Heavenly Spheres), the first copy of which reached him on his deathbed.[14]

In the Copernicus model, the planets circled the Sun; the crucial sentence in his book is "In medio uero omnium residet Sol". (But the Sun resides at the center of everything.) This contradicted the accepted geocentric model formulated in Claudius Ptolemy's *Almagest* and embraced by the Catholic Church.[15] Copernicus studied mathematics in the Polish city Cracow (Kraków) and medicine and canon law in Italy and was a versatile polyglot with many interests. He never made his living as an astronomer; he was a physician and the personal secretary of a bishop in the Polish town, Toruń. Copernicus also calculated, very accurately, the orbital periods of the planets around the Sun:

Mercury – 80 days, Venus – 9 months, Earth – 1 year, Mars – 2 years, Jupiter – 12 years, and Saturn – 30 years.

The page from *De revolutionibus orbium coelestium* with a drawing of the solar system, with the Sun in the center, and information about the lengths of the planets' orbits.[16]

One of the erroneous assumptions made by Copernicus was that the orbits of the planets were circles, with the Sun residing in the center. It took more than 150 years to find out that this was not the case.

Tycho Brahe One of the scientists whose work eventually led to the conclusion that orbits are ellipses was a Danish nobleman, Tycho Brahe (1546–1601), who was also an astronomer and an alchemist. For more than 20 years, he made detailed astronomic measurements in his observatory on the Danish island, Hven. Eventually, he produced an impressive catalog that gave the positions of 1,000 stars.

By means of these painstaking naked-eye observations, he came to the conclusion that Ptolemy's model of the universe was incorrect. But instead of adopting the heliocentric system of Copernicus, he came up with his own 'intermediate' model of the universe, in which the Sun orbited the Earth and the other planets orbited the Sun.

Johann Kepler   *Johann Kepler* (1571–1630), an impoverished German astrologer and a former university teacher, worked for Brahe after he moved from Denmark to Prague. Kepler used Brahe's and his own observation tables to eventually modify the Copernican model and formulate his three laws of planetary motions. One of them stated that the orbits of the planets are ellipses, with the Sun residing in one of the foci. This bold conclusion is one of the most daring hypotheses put forward in the history of science; until then, the ellipse was not observed in nature. This law was confirmed only a year after his death, when Mercury was spotted passing in front of the Sun exactly as predicted.

The other two laws specified in detail the speed of the planets and compared the total orbit times of a pair of planets. These three laws were confirmed mathematically more than 150 years later.

Gravitation by M.C. Escher.[17]

Kepler was also a talented mathematician. In particular, he discovered new regular three-dimensional objects that go beyond the Platonic solids and studied their mathematical properties. His idea was that, if one drops the requirement of convexity (which means that any line connecting two points of the object remains inside it), more objects can be found. This way, he discovered so-called non-convex regular polyhedra. One of them inspired the impressive work, *Gravitation*, by the 20th-century Dutch artist Maurits Cornelis Escher.

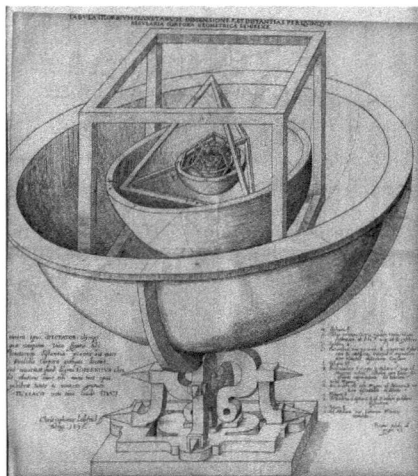

Kepler's model of universe from his book *Mysterium Cosmographicum*.[18]

This fascination with regular solids brought Kepler on a spurious track. In his book, *Mysterium Cosmographicum* (The Cosmographic Mystery), in which he praised the Copernican system, he also proposed a model of the universe in which one Platonic solid fits between each pair of planetary spheres.

This idea definitively collapsed when a new planet, Uranus, was discovered in 1781. He also published his third law of planetary motions in his opus magnum *Harmonices Mundi* (The Harmony of the World), a book on which he worked for 20 years, alongside a strange theory that tried to connect Platonic solids with the classical elements of fire, earth, air, water, and the cosmos.

One of the spectacular applications of Kepler's laws was the determination of the Sun's distance to Earth; namely, the position of the Sun could now be computed by determining the (elliptical) orbit of a single planet and subsequently computing the location of the focus of the ellipse, in which the Sun resides. As already mentioned, the first realistic computation of the Sun's distance to Earth was done in 1672, by studying the orbit of Mars. A century later, more accurate computations were obtained by studying the orbit of Venus from eight nations, during a rare event called the *transit of Venus*, when Venus passes in front of the Sun. These transits usually come in pairs eight years apart. The last two took place in 2004 and 2012 and the next one will take place in 2117.[19]

Galileo Galilei   In August 1597, the Italian astronomer **Galileo Galilei** (1564–1642) received a copy of Kepler's book *Mysterium Cosmographicum* from him. In his reply, Galileo admitted that he had also embraced the Copernican view of the universe. Twelve years later, once he had produced a telescope, Galileo could confirm the validity of the Copernican system by discovering Jupiter's moons and finding that they circled Jupiter and not the Earth. Galileo was a keen experimenter and sought to capture the workings of nature in mathematical laws. This made him one of the first modern scientists. He memorably stated: "The book of nature is written in the language of mathematics."

Galileo did not make new contributions to mathematics but knew how to make innovative use of it in astronomy and physics. Combining his observations of the Moon through his telescope with a simple reasoning based on the Pythagorean theorem, he estimated that the height of a specific mountain on the Moon exceeded 6,500 meters. Through experiments he discovered that the trajectory of a projectile was a parabola, though he was unable to prove it. This discovery led to a formula allowing one to calculate the range of a cannon.

Until his time, following Aristotle, it was believed that heavier objects fall faster than lighter ones. Galileo contested this by means of experiments. Combining them with geometric reasoning, he realized that the velocity of a freely falling body is proportional to the elapsed time, and the traveled distance is proportional to the square of the elapsed time. Not surprisingly, Albert Einstein called Galileo the 'father of modern physics'.

René Descartes   Galileo's conflict with the Catholic Church about his opinions about the universe led to his arrest in 1633. It came as a shock to **René Descartes** (1596–1650), a Frenchman who, in spite of being a devout Catholic, spent most of his adult life in the more liberal Protestant Netherlands. Descartes is considered the 'father of modern philosophy' because he embraced rationalism as the way of acquiring knowledge. But he was also a prominent mathematician and, in particular, read Galileo's works in which Galileo accepted the Copernican model of the universe.

Descartes' first publication was the famous collection of essays *Discours de la Méthode pour bien conduire sa raison, et chercher la vérité dans les sciences* (*Discourse on the method of rightly conducting one's reason and of seeking truth in the sciences*) that appeared in 1637, when he was 41 years old. The famous phrase "Je pense, donc je suis" ("I think, therefore I am") comes from this treatise.[20] In the third essay, titled *La Géometrie*, Descartes introduced what we now call *Cartesian coordinates* that associate two coordinates with each point on a plane. This and other ideas put forward in this essay connected the fields of geometry and algebra, which until then were essentially separate.

This combined field of study, called *analytic geometry*, since the 19th century became crucial for the subsequent developments in mathematics, physics, astronomy, and engineering. The reason is that it allows one to describe figures, curves, surfaces, and motion, by means of algebraic expressions, so one can, for instance, study conic sections, in particular, planetary trajectories, using algebraic methods.

In his discussion of curves, Descartes was already using a modern notation according to which constants are denoted by the first letters of the alphabet and the unknowns by the last ones, while the powers are written using superscripts, for example, x^3 to denote x raised to the third power.

Napier and Briggs   Another contribution in this period revolutionized computing. An apparently quarrelsome Scot[21] **John Napier** (1550–1617) published in 1614 a book that was the culmination of his 20 years of research into the concept of a *logarithm*. Napier produced tables of logarithms that allowed him to compute multiplications and divisions of large numbers by means of the much simpler operations of addition and subtraction. A couple of years after his death, this idea was improved upon by the first professor of geometry at the University of Oxford, **Henry Briggs** (1556–1630), who produced in 1624, tables of logarithms with base 10. This made computations using logarithms much simpler. These matters are explained in Appendix 18.

Eventually, Briggs produced logarithms for all numbers up to 100,000. Soon afterward, his tables became a standard tool in astronomic calculations. In particular, Kepler used them.

Nowadays, the *logarithmic scale* is used to represent in a convenient form wide-ranging quantities, by translating *orders of magnitude* to a more convenient linear scale. Examples are the *decibel* scale, which measures the loudness of sound, and the *Richter scale*, which measures the strength of earthquakes. Consequently, an increase of 1 on the Richter scale, say from 4.5 to 5.5, means a tenfold increase in the amplitude.

Logarithms are also used in determining the age of living organisms. This is done by means of the ingenious *radiocarbon dating* method proposed by Willard Frank Libby, an American chemist who won the Nobel Prize in Chemistry for it in 1960. This method measures the radioactive decay in terms of *half-life*, the time it takes for half of the nuclei sample to decay.[22] A famous example of its use was determining the age of Ötzi the Iceman, a mummy from the Neolithic period, discovered in 1991 in the Alps. The results showed that he died over 5,000 years ago.

Foundations of probability   The period I am discussing led also to the rise of *probability*, a field concerned with the concepts of chance and risk. As already mentioned, Cardano was a compulsive gambler, which led him to study the chances of winning in card games and games of dice. No wonder then that he wrote the first treatise on probability. But the proper foundations of the field crystallized later through a correspondence between two brilliant Frenchmen, **Blaise Pascal** (1623–1662) and **Pierre de Fermat** (1601–1665). They repeatedly tried but never met because of Pascal's protracted sickness and premature death due to a brain tumor and stomach cancer.

Pascal had been a child prodigy and was an ascetic polymath. To help his father, a tax collector, he produced at the age of 21 several copies of a calculating machine, now called the *Pascaline*, which performed additions and subtractions. It is a precursor to modern calculators.

In his remarkable book *Pensées* (Thoughts) that he wrote in the painful months preceding his death, Pascal used the following probabilistic argument, now called *Pascal's Wager*, to justify his belief in God:

> Belief is a wise wager. Granted that faith cannot be proved, what harm will come to you if you gamble on its truth and it proves false? If you gain, you gain all; if you lose, you lose nothing. Wager, then, without hesitation, that He exists.

Fermat was a lawyer and government official. Even though he considered mathematics a hobby, this did not prevent him from being one of the leading mathematicians of his time. He came up with a method of computing the maximum and the minimum of a function and successfully tested it on examples such as the intuitive theorem of Euclid, which states that among all rectangles with a given perimeter, a square has the largest area. This formed the beginning of *differential calculus*, the subject of which is the study of the rate of change of various quantities, for instance, speed.

Also, Fermat postulated that light travels between two given points (for example, one located in the air and the other in water) along the path of shortest time. This idea eventually led to the fundamental principle of *least action* widely used in physics.

Further, he is considered to be the 'father of modern number theory' in which, in particular, he established a classic result about prime numbers called *Fermat's little theorem*.[23] Fermat, independently of Descartes, also connected algebra with geometry. Nowadays, they are viewed as the founding fathers of the field of analytic geometry. However, Fermat is most famous for a result that he most probably did not prove and which I shall discuss in the chapter on the 20th and 21st centuries.

Isaac Newton

An often-made remark is that the year Galileo died, *Isaac Newton* (1642–1727) was born, but this is true only because England adopted the Gregorian calendar only after Newton's death. Newton was the son of an illiterate English farmer and rose to such fame that he was granted a state funeral and buried in Westminster Abbey. He was a modest and solitary person, albeit prickly, and as we shall see in one infamous example, easily prone to conflicts with other scientists. The famous statement "If I have seen further, it is by standing on the shoulders of giants" is ascribed to him.

Newton's career was strikingly fast. At the age of 27, he became a professor of mathematics and physics at the University of Cambridge, just a year after he attained there his MA. He held this post for the next 27 years. Afterward, he moved to London, where, from 1699 until his death, he was the Master of the Mint, then an important office in the English government. In this capacity, he oversaw the production of coins and vigorously pursued counterfeiters. Moreover, from 1703, until his passing, he was also the president of the Royal Society, over which he presided in a dedicated, though dictatorial, manner. During his life, he traveled very little and, in particular, never left England.

In his 1687 three-volume book *Philosophiae Naturalis Principia Mathematica* (The Mathematical Principles of Natural Philosophy)—usually referred to simply as *Principia*)—that he wrote in just 18 months, he essentially created a framework for classical mechanics by introducing the notions of force, inertia, and momentum, and by formulating three laws of motion that relate them. In the book, he also described the working of the solar system and, among others, derived mathematically Kepler's laws from his own laws of motion and law of universal gravitation.[24]

The progression from the Copernican model to the detailed observations of Brahe that he carried out over the period of 20 years, to Kepler's formulations of the laws about the planet's elliptic orbits, and finally to Newton's proofs of these laws took almost 140 years and is a mind-boggling triumph of human intellect. This story inspired Alfred Noyes, an early 20th-century English poet, to write the now forgotten and more than 100-page-long poem *Watchers Of The Sky*, devoted to Copernicus, Brahe, Kepler, Galileo, Newton, and two other astronomers.[25]

Newton also showed that his law of universal gravitation could always predict the orbit of two bodies held together by gravity, for instance, the Sun and the Earth. But he was unable to tackle the problem of three bodies, for instance, the Sun, the Earth, and the Moon. This problem is notoriously difficult and, until today, only partial solutions have been found. I shall return to the matter when discussing the mathematics of the 19th century.

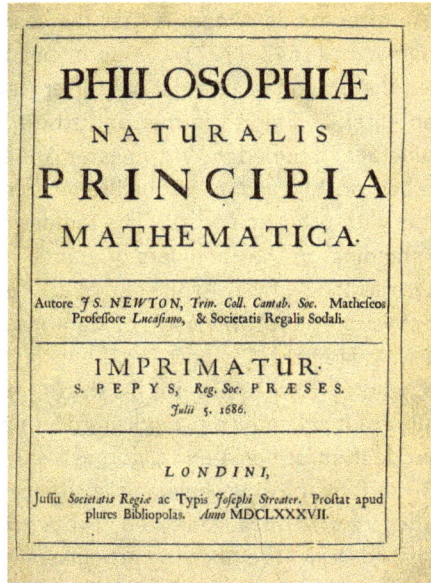

Title page of the first edition of *Principia* (1687).[26]

As an aside, the famous anecdote that Newton came up with his theory of gravity after he was struck by an apple falling from a tree seems to have been invented by Newton at an old age.[27]

Newton also, incredibly, correctly predicted that the Earth is not completely round because the rotation force caused it to flatten at its poles. Fifty years later, this fact was confirmed by means of tortuous triangulations carried out in dramatic conditions by two teams of French scientists in Finland and Peru.[28] Newton is responsible for many other fundamental contributions to mathematics, and a number of mathematical concepts and methods bear his name. Most importantly, he created *calculus*. Morris Kline, a prominent historian of mathematics, considered calculus the greatest creation in mathematics, next to Euclidean geometry.[29]

The following quotation from Newton's letters shows his influence by Fermat ('fluxion' is Newton's name for the rate of change, the subject of what we now call differential calculus):[30]

> I had the hint of this method [of fluxions] from Fermat's way of drawing tangents, and by applying it to abstract equations, directly and invertedly, I made it general.

The main revolutionary innovation of calculus is that—assuming we are concerned with the computation of an area under a curve—Archimedes' laborious analysis based on successive sequences of approximations can be replaced by a single operation that associates with a given function an *integral* (more precisely an *indefinite integral*), a function that can then be used to compute the area in a straightforward way. So the whole process of computing the area is essentially reduced to finding an integral of an appropriate function. This can be done systematically for several (though not all) functions arising in practical applications, a process started by Newton and independently by Leibniz. Finding an integral of a complicated function can be usually done by combining (sometimes in a clever way) some rules with consulting a table of integrals of some basic functions. In fact, books on mathematics for engineers routinely contain such a table on the inner cover. Nowadays, finding an integral of a function can be done by using appropriate free online computer systems. I shall discuss one such system in Chapter 8.

However, making the foundations of calculus rigorous was another matter. In 1734, the Irish philosopher Bishop George Berkeley published a pamphlet highly critical of Newton's use of 'infinitesimal quantities', which he contemptuously called "the ghosts of departed quantities". Such well-founded reservations were only dissipated in the 19th century when, thanks to a sustained effort of the most prominent mathematicians, the fundamental notions of calculus were finally properly formalized.

Gottfried Wilhelm Leibniz Newton published his findings very reluctantly. The same was the case with a German, **Gottfried Wilhelm Leibniz** (1646–1716), who independently invented calculus as well. Leibniz was a child prodigy. When he was eight years old, he began to teach himself Latin and, a couple of years later, started to study Greek. During his childhood, he was an avid reader of classics. At the age of 15, he entered University and got his doctorate in law five years later. In his twenties, he constructed a calculating machine that was more advanced than Pascal's Pascaline since it could perform addition, subtraction, multiplication, and division. In 1673, this earned him a membership of the Royal Society of London. Leibniz was one of the most respected scientists of his time. In 1700, he became the president of the newly founded Berlin Academy of Sciences, a title he held until the end of his life.

The protracted and acrimonious dispute between two intellectual giants, Newton and Leibniz, about the priority of their discovery of calculus is one of the less proud pages in the history of mathematics. Newton came up with his approach to calculus in 1665–1666 (he was only 22–23 years old then!) but published his findings only in 1704. Leibniz developed his own approach in 1673–1676 (he was 27–30 years old then) and did not rush to publish his findings either. Still, he published his work earlier than Newton, in the period 1684–1686. In the preceding 20 years, he had occasional correspondence with Newton (they never met) and had some cursory insight into Newton's work through other published works. Over the years their conflict slowly brewed, often instigated by other mathematicians. In 1711, Leibniz made a foolish decision to appeal against the charges of plagiarism to the Royal Society, of which Newton was the president. The Society unsurprisingly sided with Newton, as he was one of the authors of the final anonymous report. Its harsh verdict, publicized during a special public meeting of the Society (convened by Newton), considerably damaged Leibniz's reputation.[31]

The current notation, the name 'calculus' (originally *calculus differentialis* and *calculus integralis*), and the so-called *fundamental theorem of calculus*[32] are due to Leibniz. Using calculus, Leibniz found, in particular, the following elegant (though hardly useful) formula for computing π:

$$\frac{\pi}{4} = 1 - \frac{1}{3} + \frac{1}{5} - \frac{1}{7} + \frac{1}{9} - \ldots$$

that links the number π with odd numbers. This formula was important because it showed that π could be defined by means of a mathematical expression.[33]

Leibniz strongly believed in the importance of logic and, throughout his life, hoped to discover a way of replacing reasoning by calculation. He was inspired by similar suggestions expressed by Ramon Llull and the philosopher Thomas Hobbes. Being more mathematically inclined, Leibniz went further.

In *Calculus ratiocinator*, he suggested a system that would perform logical deductions and started working towards it by transforming syllogisms into simple operations on sets. This was the first step towards linking logic with algebra, a far-reaching idea rediscovered and presented in a more rigorous form some 150 years later by George Boole. Leibniz's suggestion essentially calls for a mechanization of reasoning.[34] In the second half of the 20th century, this was realized in various computer programs called *automated theorem provers*.

In several manuscripts, starting from 1674, Leibniz also extensively discussed his two other inventions, also relevant for computer science. One was the *binary notation*, in which one writes numbers using only two symbols, 0 and 1, now widely used in computers. The other was the *hexadecimal notation*, which is based on 16 digits. Nowadays, it is used in various applications in computer science; for example, to encode colors in web pages.[35]

First page of a Leibniz manuscript from 1705 about binary notation.[36]

Leibniz was one of the most impressive figures in the history of science. The breadth of his scientific and philosophical interests calls for a comparison with Aristotle.[37] His boundless curiosity led him, in particular, to philosophy (he developed a theory of self-contained, simple substances that he called *monads* and was an influential proponent of the *Principle of Sufficient Reason*, asserting that nothing happens without a reason), psychology (he introduced the concept of subconsciousness), linguistics (he tried to construct a linguistic genealogy for the main languages of Europe and Asia), physics (the word 'dynamics' is due to him), history (his study of the German Empire in the Middle Ages was later used by Edward Gibbon in his monumental treatise *The History of the Decline and Fall of the Roman Empire*)[38], and to many ideas, proposals, and inventions, including a recently discovered design of a cryptographic machine.[39]

Not all his ideas were accepted. He tried to unify the Catholic and Protestant doctrines, obviously without success. Also, he was a proponent of a doctrine that we live in the best possible world. Forty years after his death, it prompted Voltaire to ridicule this idea in his most famous work, *Candide*.

Christiaan Huygens A contemporary of Newton and Leibniz was the Dutchman **Christiaan Huygens** (1629–1695), one of the leading scientists of his time. In particular, he was a prominent astronomer who, among others, discovered the rings of Saturn. Also, he proposed the wave theory of light, which implied its finite speed. After receiving crucial data of eclipses of one of Jupiter's moons, Huygens computed the speed of light (with an error of 25%). Until then, since the times of Aristotle, it was assumed that it had an infinite speed. Throughout his life, Huygens met and discussed science with the major scientific figures of his time: Pascal, Descartes, Leibniz, and Newton, and was one of the founding members of the French Academy of Sciences.

He learned of the mentioned correspondence between Pascal and Fermat, which inspired him to write the first modern book on probability. He used it in his study of life expectancy that he carried out with his brother. The following quotation provides an insight into his work on this subject:[40]

> There are thus two different concepts: the expectation or the value of the future age of a person, and the age at which he has an equal chance to survive or not. The first is for the calculation of life annuities, and the other for wagering.

We also owe to Huygens the first analysis of the risks involved in gambling in a casino. It goes nowadays under the catchy name of *gambler's ruin*. Consider a gambler who goes to a casino with some amount of money and—naturally—plans to leave with a larger amount. Each time he places a fixed small bet, he can win with some fixed probability. The question is: what are the chances of the gambler going bankrupt after a given number of rounds?

One of the early applications of calculus was a solution of the *tautochrone problem* (in Ancient Greek, 'tauto' means 'the same' and 'chronos' means 'time'). The problem, posed by Leibniz, asks to determine the shape of the curve between two points with the property that no matter where a bead is placed on it, it will slide to the lower point in the same amount of time.

Huygens found that the solution was an inversion of a *cycloid*, which is a curve formed by a point traveling on the rim of a wheel (think of the path traced by a valve of a tire on a moving bike).

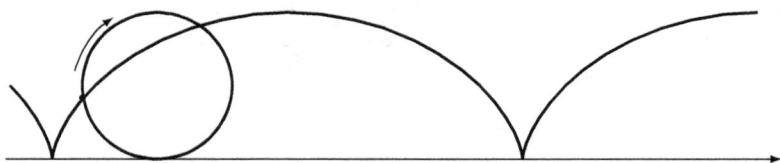

A cycloid.

Subsequently, he used it in a brilliant way in his construction of a pendulum that ensured the correct working of a clock by forcing its string to follow a cycloid. The cycloid fascinated several prominent mathematicians of the 17th century, and they discovered various beautiful properties of it.[41] Also, Pascal is known to have studied it when suffering from a severe toothache at the later stage of his short life.

In the discussed period, a number of other brilliant mathematicians made relevant though less important contributions. One of them, a well-to-do Frenchman, **Claude-Gaspar Bachet de Méziriac** (1581–1638) wrote the first ever book on puzzles and mathematical recreations, titled *Problèmes Plaisants et Délectables qui se Font par les Nombres* (Pleasant and Delectable Problems that Occur Through Numbers). It went through five editions, the last one as late as 1959. A number of classic puzzles can be traced back to this book.[42]

A complex period	The scientific revolution led to dramatic changes in mathematics, physics, and astronomy, and also in

other sciences, notably in biology and chemistry. Yet, the surrounding circumstances show how complex this period was. Copernicus, in order not to anger the Catholic Church, delayed the publication of his seminal book *De revolutionibus orbium coelestium* almost until his death.

Life did not spare Mercator, who spent seven months in prison, accused of heresy.[43] In turn, Napier was most proud of his widely read polemical tract in which, among others, he claimed that the Creator proposed to end the world between 1688 and 1700.[44] Further, Briggs published an influential small tract in which he claimed that California was an island.

Galileo, as a result of the process carried out by the Inquisition, had to recant his opinions and was confined to house arrest until the end of his life. Incidentally, he was teaching mathematics and (Ptolemaic) astronomy at the University of Padova to help students become competent in the preparation of the astrological charts that were used to properly treat various illnesses.[45]

Kepler partly stopped his work to concentrate on a protracted (and eventually successful) fight to prove that his mother was not a witch. The works of Copernicus, Galileo, and Kepler (though strangely not of Newton) ended up on the Index of Prohibited Books set up by the Catholic Church and remained on it until 1815.[46]

Newton also published several works on the occult, in particular, on alchemy. John Maynard Keynes, a leading economist of the 20th century (also an author of a well-received book on probability theory), purchased and extensively studied Newton's archives. In a lecture he prepared for the celebration of Newton's tercentenary anniversary, he quipped that "Newton was not the first of the age of reason, he was the last of the magicians [...]."[47]

Newton also wrote the book *Chronology of Ancient Kingdoms Amended*, which presented a drastically revised timeline for ancient civilizations, shortening Greek history by 500 years and Egypt's by a millennium.[48] This has given ammunition to those questioning the accepted timeline even until today.[49]

Another interesting case is Christiaan Huygens, who devoted the last ten years of his life to writing a book in which he presented various 'probable conjectures' about life on other planets. In particular, he argued that the inhabitants of Jupiter and Saturn have the incentive to study astronomy because of the multiple moons these planets have. The book stirred peoples' imagination and was a huge success. In particular, Thomas Jefferson (to whom I shall return) had a copy in his library and Peter the Great ordered its translation into Russian.[50]

In this list of curious facts, even Leibniz can be mentioned. In his proposal of binary numbers, he identified 1 with God and 0 with the void and sent this suggestion to the Jesuit president of the Chinese tribunal for mathematics for use as an argument to convert the Chinese emperor to Christianity.[51]

Timeline

Piero della Francesca (c. 1415–1492)
Luca Pacioli (c. 1447–1517)
Scipione dal Ferro (1465–1526)
Albrecht Dürer (1471–1528)
Nicolaus Copernicus (1473–1543)
Niccolo Tartaglia (1499/1500–1557)
Gerolamo Cardano (1501–1576)
Gemma Frisius (1508–1555) (mentioned in Chapter 2)
Gerardus Mercator (1512–1594)
Lodovico Ferrari (1522–1565)
Raphael Bombelli (1526–1572)
François Viète (1540–1603)
Tycho Brahe (1546–1601)
Simon Stevin (1548–1620)
John Napier (1550–1617)
Henry Briggs (1556–1630)
Galileo Galilei (1564–1642)
Johann Kepler (1571–1630)
Willebrord Snellius (1580–1626) (mentioned in Chapter 2)
Claude-Gaspar Bachet de Méziriac (1581–1638)
René Descartes (1596–1650)
Pierre de Fermat (1601–1665)
Blaise Pascal (1623–1662)
Christiaan Huygens (1629–1695)

Isaac Newton (1642–1727)
Gottfried Wilhelm Leibniz (1646–1716)

Notes

[1] This formula is as follows. Let $(a, x), (b, y), (c, z)$ be the six edges of the tetrahedron grouped into opposite pairs.

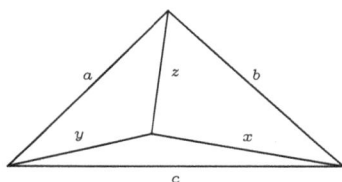

A tetrahedron.

Then

$$144V^2 = -a^2b^2c^2 - a^2y^2z^2 - b^2x^2z^2 - c^2x^2y^2$$
$$+a^2c^2z^2 + b^2c^2z^2 + a^2b^2y^2 + b^2c^2y^2$$
$$+b^2y^2z^2 + c^2y^2z^2 + a^2b^2x^2 + a^2c^2x^2$$
$$+a^2x^2z^2 + c^2x^2z^2 + a^2x^2y^2 + b^2x^2y^2$$
$$-c^2c^2z^2 - c^2z^2z^2 - b^2b^2y^2 - b^2y^2y^2$$
$$-a^2a^2x^2 - a^2x^2x^2,$$

where V is the volume of the tetrahedron considered.

[2] Courtesy: Public Domain, Wikipedia Commons.

[3] The controversy around solving third-degree equations is discussed in detail in M. Livio, *The Equation That Couldn't be Solved*, Simon & Schuster, pp. 63–78, 2005.

[4] Omar Khayyam expressed such equations as an intersection of a hyperbola and a parabola, see G.G. Joseph, op. cit., pp. 482–484.

[5] A close friend of Pascal questioned that $\frac{-1}{1} = \frac{1}{-1}$ using a seemingly convincing argument that—using modern notation—if $a < b$ (here $-1 < 1$), then $\frac{a}{b}$ cannot be equal to $\frac{b}{a}$, see M. Kline, *Mathematical Thought From Ancient to Modern Times*, Oxford University Press, p. 252, 1972.

[6] His rules were as follows:

piu di meno via piu di meno fa meno (i times i makes less),
piu di meno via men di meno fa piu (i times $-i$ makes more),
meno di meno via piu di meno fa piu ($-i$ times i makes more),
meno di meno via men di meno fa meno ($-i$ times $-i$ makes less).

This makes perfect sense if we remember that i stands for $\sqrt{-1}$ and interpret 'less' as -1 and 'more' as 1. For example, the first rule states that $\sqrt{-1} \cdot \sqrt{-1} = -1$.

[7] This representation was independently introduced by C. Wessel in 1799 and by J.-R. Argand in 1813 and was also known to C.F. Gauss discussed in Chapter 7. It is now called the *Argand diagram*.

[8] The first such use is due to A.-J. Fresnel, who, in 1823, applied complex numbers to his study of refraction and reflection of polarized light, see R. Karam, Fresnel's original interpretation of complex numbers in 19th century optics, *American Journal of Physics*,

86(4), pp. 245–249, 2018.

[9]I. Grattan-Guinness, op. cit., p. 189.

[10]L. Hogben, *Mathematics for the Million*, George Allen & Unwin Ltd, p. 303, 1940. I inherited this book from my father.

[11]F.Q. Gouvêa, op. cit., p. 80. To be more precise, Stevin used a bit more complicated notation. The current notation was introduced by John Napier.

[12]Courtesy: Public Domain, Wikipedia Commons.

[13]Courtesy: Public Domain, Wikipedia Commons.

[14]J. Bronowski and B. Mazlish, *The Western Intellectual Tradition*, Dorset Press, p. 113, 1960.

[15]The Sun-centered model was proposed in Ancient Greece by the already mentioned Aristarchus of Samos, but his views on this matter were received with hostility and subsequently ignored.

[16]Courtesy: Public Domain, Wikipedia Commons.

[17]Courtesy: Public Domain, Wikipedia Commons.

[18]Courtesy: Public Domain, Wikipedia Commons.

[19]A.W. Hirshfeld, op. cit., pp. 61–65.

[20]The more known Latin wording "Cogito ergo sum" appeared in his *Principia Philosophiae* (Principles of Philosophy), published in 1644.

[21]E. Maor, *e: The Story of a Number*, Princeton University Press, p. 4, 2009.

[22]This method makes use of the fact that the amount of the radioactive carbon-14 in the air has stayed unchanged for thousands of years. Once an organism dies, its absorption of carbon-14 through breathing or eating stops. From this moment on, radioactive decay takes place. By comparing the ratio of carbon-14 in the found organism with that in the atmosphere, one can estimate the age.

[23]This result states that for any natural number n, if p is a prime number, then it divides $n^p - n$.

[24]Newton's three laws of motion are:

First law: A body that is at rest (or in motion) will remain so until acted upon by an external force.

Second law: The change in motion is proportional to the applied force.

Third law: To every action there is always an equal opposite reaction.

His law of universal gravitation expresses the gravitational force between two bodies in terms of their masses m_1 and m_2 and their distance r. This force equals $G\frac{m_1 \cdot m_2}{r^2}$, where G is the *universal gravitational constant*.

[25]Here is a typical fragment:

When, in the year of Galileo's death,
Newton, the mightiest of the sons of light,
Was born to lift the splendour of this torch
And carry it, as I heard that Tycho said
Long since to Kepler, "carry it out of sight,
Into the great new age I must not know,
Into the great new realm I must not tread."

[26]Courtesy: Public Domain, Wikipedia Commons.

[27]S. Connor, The core of truth behind Sir Isaac Newton's apple, *Independent*, 18 January 2010, http://www.independent.co.uk/news/science/the-core-of-truth-behind-sir-isaac-newtons-apple-1870915.html.

[28]E. Danson, op. cit.

[29]M. Kline, op. cit., p. 342.

[30] A.I. Sabra, *Theories of Light: From Descartes to Newton*, Cambridge University Press. p. 144, 1981.

[31] H. Hellman, *Great Feuds in Mathematics*, Wiley, pp. 51–72, 2006.

[32] This theorem relates the concepts of the (indefinite and definite) integral and the derivative.

[33] This formula was also known to a 14th-century Indian mathematician Madhavan and a contemporary of Leibniz, a Scottish mathematician, J. Gregory.

[34] J. Gray, op. cit.

[35] See L. Strickland and H. Lewis, *Leibniz on Binary: The Invention of Computer Arithmetic*, The MIT Press, 2022.

[36] Courtesy: Public Domain, Wikipedia Commons.

[37] A. Gottlieb states plainly in his book *The Dream of Enlightenment*, Allen Lane, p. 163, 2016 that "Leibniz [...] was the greatest polymath since Aristotle; there has not yet been a third person who can stand alongside them."

[38] G.F. Simmons, *Calculus Gems, Brief Lives and Memorable Mathematics*, The Mathematical Association of America, pp. 141–157, 2007.

[39] A. Gottlieb, op. cit., pp. 163–195.

[40] The quote is from A. Hald, *A History of Probability and Statistics and Their Applications before 1750*, Wiley-Interscience, p. 106, 2003.

[41] In particular, G. de Roberval and E. Torricelli, the inventor of the barometer, independently found that the area under the fragment resulting from one revolution of the circle equals three times its area, while Ch. Wren, the architect of St. Paul's Cathedral in London, found that the length of this fragment is four times the diameter of the circle.

[42] Here is one example. Find a system of four weights that adds up to 40 and allows one to weigh using a scale every integral weight from 1 to 40. The answer is 1, 3, 9, 27. For example, to get the weight 20, we put on one scale 27 and 3 and on the other 9 and 1.

[43] E. Danson, op. cit., p. 11.

[44] D.M. Burton, *The History of Mathematics: An Introduction*, McGraw-Hill Science, 7th edition, p. 351, 2011.

[45] I. Grattan-Guinness, op. cit., p. 177.

[46] Remarkably, even in 1993, Cardinal Ratzinger, who became Pope Benedict XVI in 2005, stated that the outcome of Galileo's trial was 'reasonable and just', see N. Pasachoff and J. Pasachoff, Galileo Galilei, in: A. Robinson (ed.), *The Scientists*, Thames and Hudson, p. 39, 2012.

[47] The symposium was delayed because of the Second World War and took place only in 1946. Keynes passed away three months before the event took place. The lecture was delivered by his brother.

[48] J.Z. Buchwald and M. Feingold, *Newton and the Origin of Civilization*, Princeton University Press, 2012.

[49] An example is the drastically revised timeline extensively discussed in a recent series of several volumes of *History: Fiction Or Science?*, Delamere Publishing, written by a prominent Russian mathematician A.T. Fomenko with coauthors.

[50] G. Basalla, *Civilized Life in the Universe: Scientists on Intelligent Extraterrestrials*, Oxford University Press, pp. 41–43, 2006.

[51] M. Kline, *Mathematics and the Physical World*, Ty Crowell Co., p. 40, 1959. Republished by Dover Publications in 1981.

Chapter 6

The 18th Century

The scientific contributions of Newton and Leibniz, together with new ideas about the role of government, religious tolerance, and the importance of empiricism in science advanced by English and French philosophers, marked the beginning of the Enlightenment. This intellectual period continued throughout the 18th century until the French revolution. It was characterized by rationalism and the rejection of traditional views on religion and political power. In mathematics, it was dominated for decades by the members of the remarkable Swiss family of Bernoulli of no less than eight famous mathematicians and their student Leonhard Euler.

Jacob Bernoulli   The most senior mathematician in the Bernoulli family was **Jacob Bernoulli** (1654–1705), a contemporary of Newton and Leibniz, though his most significant contributions—in probability—were published in the 18th century. Bernoulli also invented the ubiquitous constant e, which together with π, is undeniably the most important mathematical constant. It naturally arises in the analysis of compound interest and in calculus.[1] His work on probability was collected in the book *Ars Conjectandi* (The Art of Conjecture), which was published posthumously in 1713. One important idea of Bernoulli, bearing his name, can be easily explained.

Statistics show that in the past years worldwide, there have been about 105 boys born for every 100 girls. We may ask: what is the chance that among 1,000 new borns, there are precisely 600 boys? Another example: what is the probability that in 20 tosses of a dice, 6 appears at least three times in a row?

The common factor in both situations is that in each single event (a birth or a toss of a dice), exactly two outcomes are possible (boy/girl or 6/not 6), each with a fixed probability. Furthermore, each event (here, a birth of a child or a toss of a dice) is independent of other ones, and we analyze its repetition several times. Such a situation is now called a *Bernoulli trial*. Clearly, this setup is common in many situations. Bernoulli came up with a formula allowing us to compute the answers to such problems in which he relied on a formula of Newton's that was concerned with combinatorics.

In his book, Bernoulli also stated that "Sometimes the stupidest man [. . .] knows for sure that the more observations of this sort that are taken, the less the danger will be of straying from the mark." He then made a first attempt to formalize and justify this intuition. This is now called the *law of large numbers*, and its most general form is a culmination of the efforts of a number of mathematicians. This law provides a justification to the colorfully named 20th-century *Monte Carlo* method that aims to obtain numerical results by repeated random sampling. It is widely used in physics, engineering, finance, and computer science. An example is the determination of the size of an oil spill that cannot be measured precisely or an assessment of the value of a complex financial portfolio.

Probability theory can easily be misused if one (often implicitly) violates some of the assumptions, for instance, that the considered events are independent, or if one interprets the problem the wrong way. For example, probability of having a boy is not fixed—it depends on a number of factors and appears to decline with the mother's age. So, we cannot blindly use the above $105/100$ ratio. In turn, a dice can be loaded. A number of famous mathematicians, including Newton and Leibniz, made basic mistakes when reasoning about simple problems involving probability.

Further, some outcomes easily defy intuition. This explains why there are so many paradoxes associated with the probability theory. One example is the *birthday paradox* proposed in the first half of the 20th century. The question is: what is the probability that in a random group of 23 people, two of them share the same birthday? It has a surprising answer: more than 50%, see Appendix 24.

Also, we seem to misjudge probabilities when reasoning about recurring events. Suppose, for instance, one tosses a fair coin 100 times. Then, the probability that one gets the same outcome at least five times in a row is surprisingly high—it is more than 97%.[2] Two more paradoxes concerned with probability will be discussed later.

| Abraham de Moivre | The approach proposed by Jacob Bernoulli was substantially generalized some 20 years later by ***Abraham de Moivre*** (1667–1754). He was born in France in a Protestant family and moved to England as a young man, after having been imprisoned for his

religious beliefs in the aftermath of the revocation of the Edict of Nantes by Louis XIV. De Moivre got interested in probability after studying Huygens' book on the subject. He also learned calculus by studying Newton's *Principia*. This brought him the idea of connecting the two subjects and to generalize Bernoulli trials to infinite situations. It led to what is now called a *normal distribution*, the familiar *bell curve* frequently seen in various statistics diagrams.

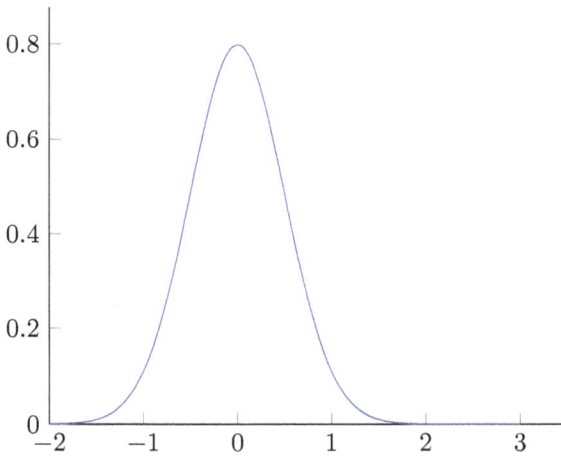

A normal distribution.

The first application of this idea was by de Moivre himself. He used the normal distribution to study mortality statistics and subsequently proposed a formula for the annuities that is a precursor of the ones used nowadays in life insurance. Also, de Moivre established the first important results on complex numbers, a concept we first encountered in the context of solving equations of the third and fourth degree, by relating them to the familiar trigonometric sine and cosine functions. The acceptance of complex numbers in the mathematical community can be traced back to his results.

Newton and de Moivre held each other in high esteem. When Newton was queried about *Principia*, he replied: "Go to Mr de Moivre; he knows these things better than I do." So it comes as no surprise that de Moivre was appointed as a member of the commission that was to judge in the dispute about the supposed plagiarism by Leibniz of Newton's work on calculus. Yet, he never succeeded in gaining a university post and eventually died in poverty.

Daniel Bernoulli   Another, highly innovative, contribution to probability was made by **Daniel Bernoulli** (1700–1782), the
nephew of Jacob Bernoulli. He analyzed what is now called the *Saint Petersburg paradox*, since his paper on the subject appeared in the proceedings of the
Saint Petersburg Academy of Sciences. It is a version of Huygens' gambler's
ruin problem pushed to infinity. It studies the situation in which a gambler
never quits, and each time he loses the bet, he doubles the stake. The paradox
is concerned with the concept of risk and is now helpful in an analysis of the
profits of insurance companies.

Thomas Bayes   Another direction in the probability theory was taken by
the English reverend **Thomas Bayes** (1702–1761). In the
later stages of his life, he got interested in probability. In his notes, which were
published posthumously, he proposed a formula, now called *Bayes' theorem*,
which allows us to update the probabilities in view of new information. This
theorem is useful in many situations in which one needs to assess statistical
evidence. For example, it was used in 1997 in court to sentence a suspect for
rape.[3] Bayes' theorem also allows us to deal with the problem of *false positives*,
tests that falsely indicate some property; for instance, a specific disease.[4] It is
discussed in Appendix 27.

A more entertaining example of its application is the following famous
Monty Hall problem.

> Imagine a stage with three doors. Behind one of them, there
> is a prize, say a car, and behind each of the other two, a goat.
> The participant is asked to pick a door. Say he chooses door
> 1. Subsequently, the show host opens another door, say door 2,
> revealing a goat. The participant is offered to reconsider his choice.
> The question is: should he switch to door 3?

This problem was conceived in 1975 but gained notoriety only after it was
published in 1990 by an American mathematician, Marilyn vos Savant, in her
weekly column in a magazine. She explained that the participant should switch,
as his chances of winning a car then increased from $\frac{1}{3}$ to $\frac{2}{3}$. Her answer
attracted many furious responses from more than 10,000 readers, including
several professors of mathematics, the majority of whom disagreed with her.[5]

Using Bayes' theorem, one can show that vos Savant was right. In this case,
the new information is that behind the door opened by the show host, there is
a goat.[6] In Appendices 28 and 29, I discuss two solutions of the problem: an
informal one and one based on Bayes' theorem.

| Leonhard Euler | The 18th century also featured a remarkable genius, the |

Leonhard Euler   The 18th century also featured a remarkable genius, the Swiss **Leonhard Euler** (1707–1783), the most prolific mathematician in history. Euler published over 30,000 pages of work on mathematics and physics. His mathematical output was only recently systematized and comprises 74 volumes.[7] Euler was equally prolific as a father. He had 13 children (though only five survived childhood) and 26 grandchildren. His photographic memory was prodigious; for example, he could recite the 10,000-line-long *Aeneid* of Virgil by heart.

Euler worked twice at the newly founded Saint Petersburg Academy of Sciences, the second time at the invitation of Catherine the Great, and at the Berlin Academy of Sciences at the invitation of Frederick the Great. This illustrates the change in attitude toward scientists during the Enlightenment and the great respect these two powerful European rulers had for Euler.

Euler made so many crucial contributions that at least three mathematical formulas are called *Euler's formula*. He contributed to geometry, calculus, number theory, algebra, and probability. In particular, he generalized Fermat's little theorem about prime numbers to a version that is now used in the RSA cryptosystem mentioned on page 17.[8]

A famous problem studied by Euler concerned seven bridges in the city of Königsberg in East Prussia (now Kaliningrad, a city lying in a Russian enclave between Poland and Lithuania). The question he posed was whether it was possible to go on a walk through all the bridges that returns to the starting point and in which each bridge is traversed exactly once.

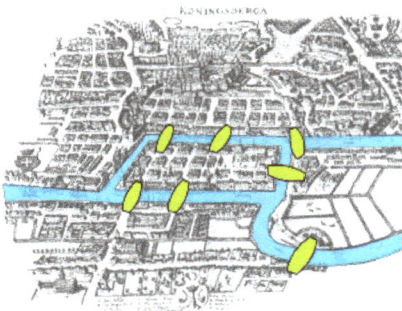

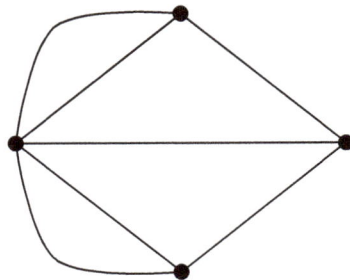

Königsberg's bridges and their representation as a graph.[9]

This is traditionally considered to be the beginning of *graph theory*, which is concerned with *graphs*, objects that consist of a set of vertices connected by edges. The figure above shows a map with the Königsberg bridges and its representation as a graph, in which each land area is depicted as a vertex and each bridge as an edge. In mathematical terms, Euler's question asks about the existence of an *Eulerian cycle* in this graph. Euler proved by a simple argument that such a walk does not exist.[10]

More importantly, he proposed a simple formula that links the number of vertices, edges, and regions in a graph that is *connected* (which means that each vertex can be reached, perhaps indirectly, from any other one) and *planar* (which means that it can be drawn on a plane in such a way that no two edges cross). Using it, one can easily establish that there are only five Platonic solids. These matters are discussed in Appendices 19 and 21.

This formula is also considered to be the beginning of *topology*, an area of mathematics concerned with the analysis of the 'shape' (in Greek, 'topos' means an object or a space) properties of the objects. Informally, topology is sometimes described as a 'rubber-sheet geometry', because it allows one to identify objects that can be transformed by stretching and contracting like rubber, so without breaking. So, no distinction is made between objects of a similar shape, for example, a doughnut and a teacup, as both objects have a single hole.

A typical example of a topological problem is that of capturing the difference between the surface of a sphere and a doughnut. The formula proposed by Euler allows one to do this by analyzing its value for planar graphs drawn on the surface of a sphere and on the surface of a doughnut. Euler's formula also allows one to solve the following well-known puzzle, first published in 1917.

> There are three houses. Each of them has to be connected to gas, water, and electricity. Prove that the connecting lines have to cross.

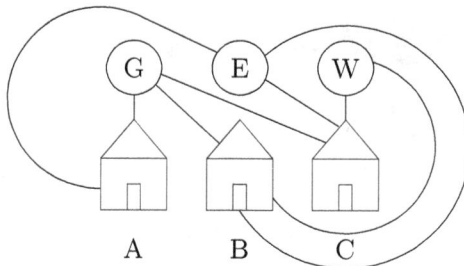

The gas, water, and electricity puzzle.

The figure above shows an incomplete solution in which house A is not connected to water (W). A solution of the puzzle using Euler's formula is given in Appendix 20.

Topology gradually became an important subject, especially during the 20th century. The 2016 Nobel Prize in Physics went to three scientists "for theoretical discoveries of topological phase transitions and topological phases of matter".

Like many mathematicians, Euler was interested in ways of computing good approximations of π. The progress in this matter can be appreciated by the fact that he came up with an unusual formula, using which he computed within an hour, a correct approximation of π to 20 decimals (compare it to the lifelong travails of van Ceulen some two centuries earlier).[11] Euler was also very influential in suggesting and promoting a notation that is used in contemporary mathematics, including e, π, i (i appeared in Chapter 5 as a symbol denoting $\sqrt{-1}$), and the notation used for the functions. He also proved that e was irrational.

Mathematicians occasionally marvel at beautiful mathematical formulas. Invariably, a stunning equation due to Euler is cited the most often. It involves exponentiation to a complex number, which needs to be defined. But one does not need to understand the details to appreciate its beauty, as it links five mathematical constants e, π, i, 0, and 1:

$$e^{i\pi} + 1 = 0.$$

There are several other topics that Euler studied, including practical problems, such as the calculation of stress in beams, the movements of the Moon, the management of seagoing vessels, and the propagation of shock waves. In spite of the fact that he was blind for the last 17 years of his life, he continued working until his last moments. Upon his death, Marquis de Condorcet commented: "*Il cessa de calculer et de vivre*" ("He ceased to calculate and to live").

Johann Bernoulli   Euler's scientific life was linked with that of the Bernoulli family. He was a student of ***Johann Bernoulli*** (1667–1748), who, in turn, was the father of Daniel and a younger brother of Jacob. Johann and Jacob shared interests in more than one mathematical subject. As a result, they were regularly involved in bitter fights about scientific priority.[12] One such topic was the infinite series, a deceptively simple subject that also caught the attention of Euler.

The earlier mentioned result of Oresme about the divergence of the harmonic numbers series

$$1 + \frac{1}{2} + \frac{1}{3} + \frac{1}{4} + \cdots$$

was lost and rediscovered more than 300 years later by Johann. It was published in 1689 by his brother Jacob who also studied the infinite sum of the reciprocals of the squares

$$1 + \frac{1}{4} + \frac{1}{9} + \frac{1}{16} + \cdots$$

Jacob proved that it converges (that is, has a finite value) but did not succeed in determining its value.

Decades after Jacob's death, Euler finally solved this problem, 90 years after it was posed. He established that this infinite sum equals $\frac{\pi^2}{6}$. Johann Bernoulli was so amazed by this result (after all, what should the number π^2 have to do with this sum?) that he exclaimed: "If only my brother was alive!"[13]

Incidentally, Jacob Bernoulli also studied the infinite sum of the reciprocals of the cubes

$$1 + \frac{1}{8} + \frac{1}{27} + \frac{1}{64} + \cdots$$

(nowadays called *Apéry's constant*) but did not succeed in evaluating it. The answer is unknown till today.

Let me conclude this diversion to Johann Bernoulli by mentioning the *brachistochrone problem* solved by him. (The word 'brachistochrone' is derived from Ancient Greek and means 'shortest time'.) It asks to determine the curve of the fastest descent between two points, one lying diagonally above the other. The problem assumes that the ball traveling on the curve has no friction and descends only due to gravity. After he found a solution, Johann challenged the readers of one of the first scientific journals to solve it. The problem was solved by his brother Jacob, Leibniz, and also Newton who found the solution overnight and submitted it anonymously.

The solution turned out to be the same as that of the tautochrone problem solved by Huygens: an inversion of a cycloid, a curve initially mentioned on page 70. The brachistochrone problem was earlier considered by Galileo, who, despite knowing and studying the cycloid (the name was given by him), wrongly concluded that the solution was an arc of a circle. But Galileo, of course, had no access to calculus.

The brachistochrone problem can be seen as the beginning of *optimal control*, an area of mathematics concerned with determining the optimal solutions subject to some natural conditions; for example, finding the minimal travel time or minimizing unemployment under some national budget constraints.

Voting and elections The second half of the 18th century produced two French mathematicians who applied mathematics to a new area: voting. This is not surprising given that they lived during the time of the French revolution. Their contributions can be seen as the first applications of mathematics in social sciences. The first of them was **Marquis de Condorcet** (in full, Marie Jean Antoine Nicolas de Caritat) (1743–1794), an influential politician and eventually a victim of the Reign of Terror. Like many French intellectuals of his time, he was deeply interested in the American Revolution and knew, among others, Benjamin Franklin and Thomas Jefferson. Condorcet held many enlightened views and, in particular, argued for women's rights and free state education.

In the voting theory, he applied the probability theory by analyzing in his *Condorcet's jury theorem*, the chances of a jury reaching a correct decision. Also, he proposed a natural method to select a winner in an election, now called a *Condorcet winner*. Unfortunately, as Condorcet himself realized, it may happen that a Condorcet winner does not exist. This complication is now called the *Condorcet paradox* and can easily arise. For example, a wine competition was organized in Paris in 1976 that went into history under the catchy name: the *Judgment of Paris*. Eleven judges had to compare ten wines through blind tasting. The results were considered shocking in France, since in both categories (white and red wines), American wines won. However, a pretty arbitrary method of computing the results was used, and a closer analysis of the outcomes showed that they suffered from the Condorcet paradox.[14]

A contemporary of Condorcet, **Jean-Charles de Borda** (1733–1799), came up with a very simple voting method, now called the *Borda count*. Suppose there are n candidates. Each voter gives n points to his/her most preferred candidate, $n-1$ points to the second most preferred candidate and so on. The winner is the candidate who gathers the largest number of points. Though this method was already used by the Roman Senate in the 2nd century, the name 'Borda count' stuck. The method has some obvious drawbacks, but it is still occasionally used.

One of the shortcomings of this method was pointed out by Condorcet in 1788, who provided an example showing that the Borda count could yield a different winner than his method. Moreover, the Borda count does not solve the Condorcet paradox. So, it is no wonder Borda and Condorcet disagreed in their views on voting. These matters are explained in Appendix 22.[15] The subject of voting was properly framed only in the second half of the 20th century when it became an area of economics based on firm mathematical grounds.

Borda deserves a place in history for another reason. Together with other

French mathematicians whom we shall soon encounter, he served on the *Bureau des Longitudes*, a committee founded in 1795 in France by the National Convention, to deal with such matters as the standardization of time and measurements. It was thanks to him that the distance of one meter was determined as one ten-millionth of the distance from the North Pole to the equator. As a result, the length of circumference of the Earth became 40,000 kilometers.

We conclude this incursion into the matters of voting by mentioning the first relevant application of mathematics coming from the United States. The previously discussed James Abram Garfield was not the first American president to be interested in mathematics. This honor should go to **Thomas Jefferson** (1743–1826), the third president of the United States, who studied mathematics and was interested in its applications throughout his life. In 1791, as the Secretary of State in the government of George Washington, he introduced the *Jefferson method* to allocate the seats in the US House of Representatives. This method tackles the common complication of fractional seats; for instance, a situation in which a political party is entitled to $12\frac{1}{3}$ seats after an election. It was used by the US Congress until 1842. The method yields the same allocation of seats as the one proposed at the end of the 19th century by the Belgian mathematician and lawyer Victor d'Hondt and which is now used during elections in several countries. The problem and the method are explained in Appendix 23.

This concludes our story of mathematics in the times of the Enlightenment. Postponed to the next chapter is a discussion of three prominent French mathematicians whose contributions took place both in the 18th and the 19th century. A remark of one of them, Pierre-Simon de Laplace, summarizes the mathematics in the 18th century succinctly in one sentence: "Read Euler, read Euler, he is the master of us all."

Timeline

Jacob Bernoulli (1654–1705)
Johann Bernoulli (1667–1748)
Abraham de Moivre (1667–1754)
Daniel Bernoulli (1700–1782)
Thomas Bayes (1702–1761)
Leonhard Euler (1707–1783)
Jean-Charles de Borda (1733–1799)
Nicolas de Condorcet (1743–1794)
Thomas Jefferson (1743–1826)

Notes

[1]The number e is defined as $\lim_{n \to \infty}(1 + \frac{1}{n})^n$. The intuition behind this expression is as follows. If a bank offers 50% (so $\frac{1}{2}$) interest paid after each half a year, then after one year, 1 unit of a currency becomes $(1 + \frac{1}{2})^2$. If it offers $\frac{1}{3}$ interest paid after each one-third of the year, then after one year, 1 unit of a currency becomes $(1 + \frac{1}{3})^3$. In general, paying an interest $\frac{1}{n}$ after each one nth of the year results in the value $(1 + \frac{1}{n})^n$. Then, e is the limit of these values.

e can be approximated to four decimals by the easy-to-remember fraction $\frac{878}{323} = 2.71826\ldots$. The importance of e in mathematics stems, among others, from the fact that it is the only constant a for which the derivative of the function a^x is again a^x. For a mathematical account of e, see E. Maor, op. cit.

[2]See, for example, S. Tijms, *Chance, Logic and Intuition: An Introduction to the Counter-Intuitive Logic of Chance*, World Scientific, 2nd edition, pp. 119–120, 2021, where several other paradoxes are discussed.

[3]P. Donnelly, Appealing statistics, *Significance*, 2(1), pp. 46–48, 2005, https://academic.oup.com/jrssig/article/2/1/46/7029499.

[4]See, for example, J.M. Almira, PCR tests with high sensitivity and specificity are truly trustworthy under high prevalence or medical prescription, *MATerials MATemàtics*, 2021(6), pp. 1–5, 2021, https://mat.uab.cat/web/matmat/wp-content/uploads/sites/23/2021/12/v2021n06.pdf.

[5]J. Tierney, Behind Monty Hall's doors: puzzle, debate and answer?, *New York Times*, 21 July 1991, http://www.nytimes.com/1991/07/21/us/behind-monty-hall-s-doors-puzzle-debate-and-answer.html. For a whole book devoted to this problem, see J. Rosenhouse, *The Monty Hall Problem: The Remarkable Story of Math's Most Contentious Brain Teaser*, Oxford University Press, 2009.

[6]It should be clarified that the show host knows behind which door the car is hidden and that he always opens a door behind which there is a goat.

[7]S. Hollingdale, *The Makers of Mathematics*, Dover Publications, p. 275, 2011.

[8]To explain his theorem, we need to introduce one notion. Given a number a, we say that b is *relatively prime* to a if a and b are commonly divisible only by 1. For example, 3 is relatively prime to 8, while 6 is not relatively prime to 8, since 6 and 8 are divisible by 2. For a given number p, Euler's function ϕ counts how many numbers smaller than p are relatively prime to it. For example, $\phi(8) = 4$, since only $1, 3, 5, 7$ are relatively prime to 8. Euler's theorem states that if p is relatively prime to n, then p divides $n^{\phi(p)} - 1$.

To see that this generalizes Fermat's little theorem, take a natural number n and a prime number p. Either p divides n—and then it also divides $n^p - n$—or p is relatively prime to n. In the latter case by Euler's theorem p divides $n^{\phi(p)} - 1$. But for a prime number p, $\phi(p) = p - 1$ because $1, 2, \ldots, p - 1$ are all relatively prime to p. So p divides $n^{p-1} - 1$. Now $n^p - n = n(n^{p-1} - 1)$; hence, in both cases, p divides $n^p - n$.

[9]Courtesy for the picture on the left: Public Domain, Wikipedia Commons. More precisely, the figure on the right represents a *multigraph* as some pairs of vertices are connected by more than one edge.

[10]His argument was based on the following observation. Consider such a walk (in an arbitrary graph). Each vertex lying on it has to be entered and left. So, for each vertex, the number of edges touching it (called the *degree of the vertex*) has to be even. However, in the graph representing the problem of the Königsberg's bridges, all four vertices have an odd degree. The requirement that each vertex has an even degree is also a sufficient condition for the existence of an Eulerian cycle.

If one relaxes the original requirement and allows the walk to end at a different place (it is then called an *Eulerian path*), then such a walk along Königsberg's bridges is still impossible. The reason is that an Eulerian path is possible precisely when zero or two vertices have an odd degree. In the first case, we even have an Eulerian cycle, and in the second case, an Eulerian path starts at one vertex with the odd degree and ends in the other. Intuitively, a graph has an Eulerian path precisely when it is possible to draw it without lifting the pencil from the paper and in such a way that each edge is traversed exactly once. A standard example of such a graph is the following drawing of an envelope:

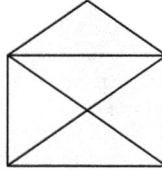

[11] The formula in question is $\pi = 20\arctan(1/7) + 8\arctan(3/79)$, where arctan is an inverse of the tan function, see W. Dunham, *The Calculus Gallery*, Princeton University Press, pp. 58–59, 2005.

[12] H. Hellman, op. cit., pp. 73–93.

[13] An exposition of Euler's proof is given in W. Dunham, *Journey Through Genius*, Penguin Books, pp. 215–217, 1991.

[14] The Judgment of Paris event is analyzed in R. Laraki and M. Balinski, How best to rank wines: majority judgment, in: E. Giraud-Héraud and M.-C. Pichery (eds.), *Wine Economics: Quantitative Studies and Empirical Applications*, Palgrave Macmillan UK, pp. 149–172, 2013.

[15] The contributions of Condorcet and Borda are extensively discussed in W. Poundstone, *Gaming the Vote: Why Elections Aren't Fair (and What We Can Do About It)*, Hill and Wang, 2009.

Chapter 7

The 19th Century

By all accounts the 19th century was a remarkable period in the history of mathematics. To quote Uta Merzbach and Carl Boyer, historians of mathematics: "The additions to the subject during these one hundred years far outweigh the total combined productivity of all preceding ages."[1]

Joseph-Louis Lagrange Some of the mathematicians who contributed to this spectacular growth of mathematics were already active in the 18th century. One of them was a largely self-taught Italian-French mathematician *Joseph-Louis Lagrange* (1736–1813), born and educated in Turin. At the tender age of 19, as the newly appointed professor at the local Royal Artillery School, he started a correspondence with Euler, who then worked in Berlin. Euler was so impressed by Lagrange's mathematical achievements that he eventually convinced Lagrange to succeed him as the Director of Mathematics at the Berlin Academy, when he decided to move to Saint Petersburg. Lagrange was only 30 then. He stayed in Berlin for the next 20 years, from where he, after the death of his patron Frederick the Great, moved to Paris.

In Paris, he was greatly honored. In particular, he became a member of the French Academy of Sciences. During the turbulent times of the French Revolution, his low political profile allowed him to avoid trouble. In 1794, he became the first professor of *analysis* (a branch of mathematics that includes calculus) at the newly founded École Polytechnique. He also served on the previously mentioned Bureau des Longitudes for some time as its chairman. Napoleon Bonaparte greatly respected him, and after Lagrange died in 1813, he was interred in the Panthéon in Paris.

Lagrange's main contributions were in mechanics, which he turned into a branch of mathematical analysis. His treatise *Mécanique Analytique* (Analytical

Mechanics), written during his stay in Berlin, summarized all the work done in this field since the time of Newton and greatly influenced subsequent work in mathematical physics. In his work, Lagrange was extremely formal and was proud that the book contained no diagrams. At the time *Mécanique Analytique* was published, Lagrange was deeply depressed by the death of his wife, and when he received the first copy from the printer, he left it unopened.

Among various results Lagrange proved in number theory, one is very easy to explain (though not easy to prove). It states that every number is a sum of at most four squares. For example, $211 = 4^2 + 5^2 + 7^2 + 11^2$. This property was originally conjectured in the early 17th century by the previously mentioned French mathematician Bachet de Méziriac, who checked it for more than 300 numbers.

Pierre-Simon de Laplace

Langrange's contemporary, **Pierre-Simon de Laplace** (1749–1827), was one of the most prominent mathematicians of his times, and he was sometimes called the French Newton. After Napoleon came to power, Laplace became the Minister of the Interior but was dismissed only six weeks later for attempting to "carry the spirit of the infinitesimal into administration".[2] Eventually, he fared better than Napoleon, and after Napoleon's fall, Laplace became a *marquis* (in English: a marquess).

Laplace was keenly interested in astronomy. In 1796, he published a popular book in which he set out to explain the origins of the solar system. The essence of his proposal was that it started as a cloud of matter (a nebula) which contracted under gravity while cooling. A similar idea had already been put forward 40 years earlier by Immanuel Kant. This theory is, therefore, now called the *Kant–Laplace nebular hypothesis*. It was rejected only at the beginning of the 20th century.

In his research on celestial mechanics, Laplace made extensive use of Newton's work in order to study the solar system. His main achievement was the five-volume treatise *Mécanique Céleste* (Celestial Mechanics) on the mathematical consequences of the law of gravitation that he published in the period 1799–1825. In it, he provided a detailed mechanical account of the solar system, which brought him to the conclusion that it was inherently stable in spite of the noted small perturbations in the orbits of planets.

Apparently, when Napoleon protested that God was not once mentioned in this work, Laplace replied: "Sire, I had no need of that hypothesis." (In contrast to Newton, who had assumed that God created the universe in 4004 BCE and who periodically intervened to preserve the stability of the solar system.)

Lagrange, after having been told about Laplace's response, commented: "Ah, but it is a beautiful hypothesis."[3]

Laplace also made fundamental contributions to probability and statistics. The main result established by him in 1810 is called the *central limit theorem* (where 'central' is supposed to mean 'fundamental'). It substantially generalized the previous result of de Moivre concerning the normal distribution because it dealt with arbitrary probabilities.

Informally, this theorem states that the averages of random observations independent of each other (think, for example, of polling potential voters) will converge to the bell curve mentioned on page 79, no matter what the original probabilities are. This remarkable result explains why the bell curve appears so routinely in statistical analysis; for example, in the distributions of income or wealth, in the statistics concerning average height, human longevity, or in the analysis of the effectiveness of a drug or specific therapy.

Two years later, Laplace published a groundbreaking treatise on probability theory. In spite of a casual remark in the introduction ("At bottom, the theory of probability is only common sense reduced to calculation"), the book consisted of 700 pages of involved mathematical analysis.[4]

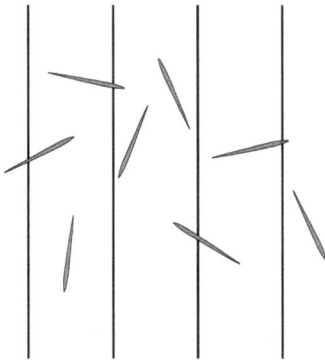

Buffon's needle problem.

One of the problems he tackled in his treatise was the *Buffon's needle problem* posed in 1777. It is a brilliantly simple idea, allowing one to determine the value of π experimentally. Imagine a floor covered with identical parallel stripes and consider a needle of length equal to the width of a stripe. Laplace proved that by repeatedly throwing the needle on the floor and counting the percentage of times it falls on a stripe, one obtains the number $2/\pi$, from which π can be calculated. Buffon's problem can be viewed as an early example of the Monte Carlo method mentioned in Chapter 6.

Joseph Fourier Like Lagrange and Laplace, ***Joseph Fourier*** (1768–1830) lived in the tumultuous times of the French Revolution and, as a student of the newly founded École Normale in Paris, was taught by both of them. His involvement in a local revolutionary committee led to his arrest and almost death under the guillotine. In 1798–1801, during Napoleon's unsuccessful military campaign in Egypt, he served as his scientific adviser. In this capacity, he was substantially involved in the only positive outcome of this campaign—a collective series of reports on ancient and modern Egypt by some 160 scientists.

A couple of years later, as a prefect of the French department of Isère, he started to work on a mathematical theory of heat. In 1807, he published his findings in an important paper *On the propagation of heat in solid bodies*, in which he expressed a function as an infinite sum of sinusoids with different frequencies, now called the *Fourier series*. Unfortunately, it was not clear when such a representation is correct and as a result, Fourier's work, called *Fourier analysis*, created a crisis in calculus. It took some 100 years to clarify the matter and find a precise answer. In spite of such initial complications, this area successfully expanded—to what is now called *harmonic analysis*—and became an important branch of mathematics. I shall return to it in the next chapter.

Fourier also was the first person who realized that the Earth's atmosphere retains heat radiation, a phenomenon we now call the *greenhouse effect*.

The mathematical contributions of Lagrange, Laplace, and Fourier were highly diverse and span a number of branches of mathematics. Yet some of these contributions share informative similarities. Each of them worked on functions, now called *transforms*, which take other functions as arguments. As a result, we talk nowadays about a *Lagrangian*, a *Laplacian*, and a *Fourier transform*. These are important examples how mathematics has been advancing through generalization, by allowing one to go beyond functions defined on numbers.

Carl Friedrich Gauss   So far, I discussed French mathematicians. But undoubtedly, the most prominent mathematician of this period was a German mathematician of poor origins, *Carl Friedrich Gauss* (1777–1855), called the Prince of the Mathematicians by his contemporaries and considered the most brilliant mathematician of all time by many. He was probably the last mathematician able to contribute to all areas of mathematics at his time. Laplace called him the first mathematician of Europe.

Gauss was, by all accounts, a child prodigy. At the age of three, he corrected an error in his father's payroll accounts.[5] At the age of 19, he showed how to construct, by using a ruler and compass, a 17-gon (a regular polygon with 17 sides), something that eluded the Greeks. In his solution, he exploited the fact, first understood by Descartes, that geometric problems can be transformed into algebraic ones and capitalized on the theory of complex numbers put forward by de Moivre, in particular, on their representation as points on a plane.

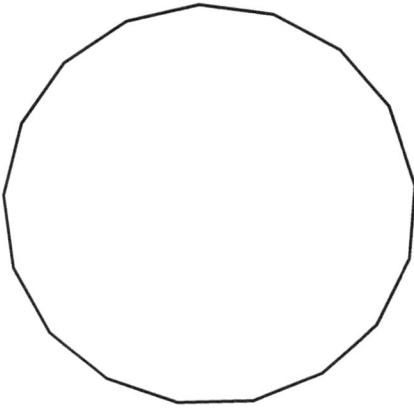

He was so proud of this result that he eventually requested that a 17-gon be engraved on his tombstone. However, after Gauss passed away, the engraver found it too daunting a task; instead a 17-pointed star was put on the base of a monument of his grave. Gauss lived most of his life in Göttingen. He hardly left the town, though he maintained an extensive correspondence with many mathematicians.

In his doctoral dissertation, he proved a classic result, now termed *the fundamental theorem of algebra*.[6] But his fame among the broader public is related to the first asteroid ever found, which was discovered on the first night of the 19th century, but was 'lost' soon after. Gauss correctly calculated its orbit from the scanty 41 days of data available, thanks to which the asteroid was 'found' almost a year later in the place predicted by his calculations. To do this, he came up with the *least squares* method used to produce a curve that best fits a set of observations that correspond to points on a plane or a space. In the 19th century, this became the main statistical method used in astronomy and geodesy.[7]

A 17-gon.

In the same year (1801), he published *Disquisitiones Arithmeticae* (Arithmetical Investigations), a groundbreaking book on number theory that he actually wrote three years earlier when he was 21. In the book, he provided foundations for *modular arithmetic*, an intuitively appealing approach to calculating 'round the clock' with an arbitrarily chosen number of 'hours'. For example, if we calculate with hours, then we compute modulo 24. So, 6 hours after 21 yields 3 o'clock, since $(21 + 6)$ divided by 24 yields the remainder 3. In modular arithmetic, one now writes this as $(21 + 6) \bmod 24 = 3$. This type of calculation is useful in many branches of mathematics, for instance, in cryptography, where it is used in the RSA cryptosystem, and in banking, where it is used in IBAN, the International Bank Account Number, to reduce transmission errors.[8]

During his life, Gauss made several highly innovative contributions to many branches of mathematics. As a result of his work on probability, the previously mentioned normal distribution (the bell curve) is sometimes called the *Gaussian distribution*. Later, I shall briefly mention his contributions to number theory and geometry. Also, in 1833, together with physics professor Wilhelm Weber, he developed the first electric telegraph, a few years before the invention of the Morse code. From the well-known quotation by Gauss, "Mathematics is the queen of the sciences and number theory is the queen of mathematics."; usually, only the first part is remembered.

The story of solving equations of fifth degree

We learned in Chapter 5 how Italian mathematicians of the 16th century succeeded in solving equations of the third and fourth degree and mentioned that the problem of solving equations of the fifth degree resisted the efforts of several mathematicians. This question was finally answered in the 19th century. The fundamental theorem of algebra proved by Gauss implies that every equation of the fifth degree with one variable, for example, $2x^5 - 5x^4 + 5 = 0$, has exactly five solutions (counting multiple occurrences, as some solutions may coincide) in complex numbers. But the question remained on how to find these solutions and how to represent them. What follows is one of the most dramatic stories in the history of mathematics.

Paolo Ruffini (1765–1822) was a brilliant Italian who was at the same time a medical doctor and a highly gifted mathematician. In 1799, he published an extremely involved proof that ran for 516 pages, showing that, in general, equations of the fifth degree cannot be *solved*. What this meant was that, unlike in the case of equations of the second, third, and fourth degree, the solutions cannot be represented as expressions that use only addition, subtraction, multiplication, and division, and the operations of extraction of the roots $\sqrt{s}, \sqrt[3]{s}, \sqrt[4]{s}, \ldots$ of an arbitrary degree of some expression s, depending only on the coefficients of the equation.

In spite of his repeated efforts (he also published two somewhat less complex proofs), no mathematician was willing to study his proofs. Several years later, gaps were found in his proofs, leaving the question of solving equations of the fifth degree unsolved.[9] The problem was finally settled by two remarkable young mathematicians; both died tragically at very young ages.

A Norwegian **Niels Henrik Abel** (1802–1829) was unaware of the work of Ruffini, and at the age of 21, established the result that Ruffini claimed, namely, that some equations of the fifth degree cannot be solved in the above sense. This crucial discovery explains why the efforts of so many mathematicians failed. Abel published the proof on his own in French, but having serious financial problems, he cut down on costs by printing only a condensed version that consisted of just six pages. He sent his proof to a number of mathematicians who ignored it. One of them was Gauss, who apparently even did not open the letter from Abel. A year later, while traveling through Europe, Abel published an extended version of his proof in a new German mathematics journal.

Throughout his short life, Abel struggled financially, and in spite of a growing recognition of his works, he could not get a permanent university position. He died of tuberculosis when he was just 26 years old. Tragically, the letter with the news that a position was finally offered to him (at the University of Berlin) arrived two days after his death.[10]

Abel settled the problem of solving equations of the fifth degree in the negative, but he did not explain how to determine that some equations can be solved (like $x^5 + x - 34 = 0$, which has 2 as a solution) while others (like $x^5 - 5x - 2 = 0$) cannot. This was clarified by a Frenchman **Évariste Galois** (1811–1832), whose life was even more dramatic and tragic than that of Abel. Galois published his first mathematical paper at the age of 18, but this was not enough to be admitted to the prestigious École Polytechnique, where he failed the entrance examination twice. He then succeeded to enter the less prestigious École Normale, where he was expelled a year later because of his political activities. In the period that followed, Galois was arrested twice and spent half a year in prison, where he tried to commit suicide.

His involvement in the republican activities against the newly proclaimed French king Louis-Philippe went hand in hand with his failed efforts to publish his groundbreaking results concerning the solvability of equations of the fifth degree, which he obtained, unaware of the works of Ruffini and Abel. One manuscript that he submitted to the French Academy was ignored, while another one was sent to Fourier, who died soon after, and so the paper was lost. His third attempt was not more successful: after six months, referees found that his proofs were unclear and too sketchy and consequently returned the manuscript.

On the eve of a duel that he was provoked to participate, he wrote a long letter to a friend, in which he explained the main lines of his theory on how to approach the problem of solving equations of the fifth degree in a systematic way and asked him to send the letter to Gauss and another German mathematician.[11] The next day Galois was dead. Apparently, his last words were "Don't cry, I need all my courage to die at 20."[12]

In his manuscripts, Galois introduced the crucial notion of a *group*. Using it, he clarified when equations in one variable of the fifth or a higher degree can be solved in the sense explained above. Galois' manuscripts were examined and eventually published only 14 years after his death. By now, his approach is a standard and classic chapter in algebra books, called *Galois theory*, while the notion of a group is one of the most successful mathematical concepts used in physics, chemistry, crystallography, cryptography, and also in the mathematical analysis of Rubik's cube.[13]

Both Abel and Galois built upon the work of Lagrange, who, in turn, relied on the work of de Moivre on complex numbers, which were discovered by Bombelli. So this line of research spans a period of more than 250 years.

Rigor in calculus   The genius of Abel and Galois was only recognized after their deaths. This was not the case with their contemporary, a Frenchman **Augustin-Louis Cauchy** (1789–1857), who, as we shall see, had to do with both of them in his career. Cauchy was a precocious child and published his first paper—on Apollonius' problem discussed in Chapter 2—at the age of 16.

He worked in a number of areas of mathematics, in which—as in the case of Euler and Gauss—various results and concepts bear his name. One of these areas is calculus. In Chapter 5, I mentioned that, from our current perspective, calculus lacked scientific rigor in the 17th century. Cauchy was one of the first in a long array of mathematicians who set out to turn it into a rigorous discipline, by properly defining the crucial notions of calculus, such as the *limit* and *continuity*. Also, he generalized calculus from real numbers to complex numbers, which led to the foundation of a new branch of mathematics called *complex analysis*. Why complex numbers are useful was summarized by another 19th-century mathematician, Jacques Hadamard, to whom I shall soon return, who once stated: "The shortest path between two truths in the real domain passes through the complex domain."

One of Cauchy's ideas, called a *Cauchy sequence*, can be easily explained. In Chapter 4, I mentioned that the infinite sum

$$1 + \frac{1}{2} + \frac{1}{4} + \frac{1}{8} + \dots$$

converges. One way to see it is by noticing that the sum (more precisely, the partial sums $1\frac{1}{2}$, $1\frac{3}{4}$, $1\frac{7}{8}$, ...) gets arbitrarily close to 2. But if the limit of a sequence (here of partial sums) is not known, for instance, in the case of the infinite sum

$$1 + \frac{1}{8} + \frac{1}{27} + \frac{1}{64} + \dots$$

studied by Jacob Bernoulli, such an argument to establish convergence does not work. Cauchy proposed instead that one should establish that the (not necessarily consecutive) elements of a sequence get arbitrarily close to each other. This approach crucially does not mention the claimed limit at all. We intuitively rely on this criterion when using infinite decimal expansions—for instance, $0.1234567891011\ldots$ (so-called *Champernowne constant*)—which, according to Cauchy's criterion, always represents a real number. This criterion is, in particular, used to argue that specific computer procedures compute a desired value in the limit.

Cauchy is also related to Abel and Galois but in a regrettably embarrassing way. Abel, while in Paris, established an important result concerning infinite sums, now called Abel's theorem and submitted a paper on it to the French Academy. Cauchy had been appointed as one of the referees, but he mislaid the manuscript and eventually forgot about it. He was also supposed to referee Galois' manuscript on the problem of solving equations of the fifth degree but never dealt with it. Cauchy was not liked by his colleague mathematicians who viewed him as arrogant. Abel wrote from Paris to a friend: "Cauchy is mad and there is nothing that can be done about him, although, right now, he is the only one who knows how mathematics should be done."[14]

Instead of refereeing, Cauchy preferred to focus on his own research. He was such a prolific author (he wrote more than 800 papers) that he started his own journal to be able to publish his manuscripts, which neatly solved the problem of refereeing them.

Another mathematician who contributed in the 19th century to mathematical rigor was the German **Peter Gustav Lejeune Dirichlet** (1805–1859), responsible for the currently used definition of a function, namely "to any x there corresponds a single y". To appreciate its simplicity consider this definition used by Johann Bernoulli a century earlier:[15]

> I call a function of a variable magnitude a quantity composed in any manner whatsoever from this variable magnitude and from constants.

Also, he was the first mathematician who found a way to repair Fourier analysis, by providing useful conditions that guarantee that a representation of a function as a Fourier series is correct. Many important contributions of Dirichlet to number theory are at the origin of *analytic number theory*, a field in which results about numbers are established by a detour through real numbers or complex numbers.

Dirichlet also came up with the intuitive *pigeonhole principle*, which states that if there are less than, say, 10 holes and if 10 pigeons fly into them, then at least two pigeons end up in the same hole.[16] A simple application of the principle is that in Amsterdam, at least two people have the same number of hair on their heads. Even though this principle sounds banal, it can be used to prove various non-trivial results; for example, every sequence of digits not starting with zero can be extended to become a power of 2. For instance, 16,777 can be extended to 16,777,216, which equals 2^{24}.

Another mathematician who crucially contributed to the rigor in calculus was a German, **Bernhard Riemann** (1826–1866), who wrote his PhD under the supervision of Gauss. His approach to computing area, volume, length of an arc, etc. (called *integration*), is now taught in the first course of calculus.

Riemann also formulated a problem about infinite sums of complex numbers, now called the *Riemann hypothesis*, widely considered to be the most important open problem in mathematics. Some deep results in number theory were established only under the assumption of this hypothesis, which explains its importance. Throughout his life, Riemann suffered from poor health and died, like Abel, of tuberculosis before he turned 40. He published only 11 papers (four more appeared posthumously), but each of them had a lasting impact.

Non-Euclidean geometry   Riemann also proposed a more general version of geometry that had far-reaching consequences. It included, as a special case, the geometry of Euclid's *Elements* and, as another special case, a four-dimensional geometry that became crucial in Einstein's general relativity theory. In another case, sometimes called Riemannian geometry, the already mentioned Euclid's 5th axiom is replaced by the one in which, given a straight line and a point not lying on it, one *cannot* draw any straight line parallel to the considered line through this point.

The first non-Euclidean geometries were actually proposed earlier by the Russian mathematician **Nikolai Lobachevsky** (1792–1856) and, independently by at his times unrecognized Hungarian genius **János Bolyai** (1802–1860), who mastered calculus at the age of 13 and spoke nine foreign languages, including Chinese and Tibetan.

After Bolyai communicated his findings to Gauss, he learned that Gauss was already aware of the existence of such alternatives to Euclidean geometry but did not want to publish this discovery out of fear that other mathematicians would reject it as being esoteric. The reply from Gauss was a severe blow to Bolyai, who never recovered from it. Lobachevsky's and Bolyai's works on non-Euclidean geometries were largely ignored until Riemann's more systematic and general approach.

Developments in statistics I already mentioned the contributions of Laplace and Gauss to statistics. It is a hotly debated discipline and a subject of many diverging opinions. Mark Twain popularized the statement: "There are three kinds of lies: lies, damned lies, and statistics", which he attributed to the 19th-century British Prime Minister Benjamin Disraeli. In turn, a 20th-century American academic, Aaron Levenstein, famously stated: "Statistics are like a bikini. What they reveal is suggestive, but what they conceal is vital."

However, there is much more to be said about statistics than such catchy remarks. It is a science that grew on top of probability theory and stands on firm mathematical foundations, thanks to three centuries of research leading to several deep and highly original results. Its dramatic progress in the 19th century is thanks to a small group of striking individuals who saw its usefulness for drawing various general conclusions about society.

A Belgian mathematician **Adolphe Quetelet** (1796–1874) learned of the work of Fourier and Laplace on probability during his visit to Paris. He came up with the idea of applying it in sociology (he termed it 'social physics'). This led him to gather and analyze various data concerning society. He became famous thanks to his book *Sur l'homme et le Développement de ses Facultés, ou Essai de Physique Sociale* (*On Man and the Development of his Faculties, or Essay on Social Physics*) on the subject, in which by focusing on characteristics such as weight and length, he discovered the importance of the bell curve in statistical analysis and also came up with the concept of an 'average man'.

Quetelet continued his analysis and found that society, on the whole, exhibits some remarkable regularities, in spite of the fact that one cannot determine anything specific about each person separately. For example, he noticed an almost constant number of crimes, such as murder, committed per year in Belgium, which captured the imagination of many people. These influential contributions to criminology brought him, in turn, to a mathematical analysis of the statistical methods he used.

He was an indefatigable proponent of applying statistics to study various aspects of society, and he encouraged several influential people in Europe and United States to gather relevant data. With this in mind, he held prodigious correspondence with some 2,500 scientists, men of letters, and monarchs.[17]

A contemporary of Quetelet was **Siméon Denis Poisson** (1781–1840), a pupil of Laplace and Lagrange at École Polytechnique, who eventually became a professor there once Fourier vacated his position. His continuing enthusiasm and devotion to mathematics are visible in his comment: "The only two good things in life are doing mathematics and teaching it." He is best known for introducing the *Poisson distribution*, a probability distribution that is different than the bell curve, which he applied to study jury deliberations in order to determine the probability of the correctness of their verdicts.

A famous application of the Poisson distribution took place during the Second World War when a British statistician used it to determine whether German bombings of London were targeting specific districts or—as he concluded—random.[18] Other examples will be provided in the next chapter. A less honorable fact to mention about Poisson is that he was one of the referees of Galois' third submission to the French Academy, which had been rejected because the proofs were found unclear.

Further progress in statistics was achieved by a remarkable Englishman, **Francis Galton** (1822–1911), a cousin and close friend of Charles Darwin. He was a child prodigy and could read and write before he was three. When he was 22, Galton inherited a substantial fortune that allowed him to pursue his personal interests for the rest of his life. In his long and fascinating life, he was, among others, a keen explorer of Africa—in particular, he tried to determine the source of the Nile—which eventually led to his election as a fellow of the Royal Geographical Society.

In statistics, he was influenced by the work of Quetelet and invented two, by now standard, concepts: *correlation analysis*, a method of measuring how a change in one variable (say, income) affects another variable (say, human longevity) and *regression analysis*, which aims to identify which variables are relevant to the study of changes of another variable. He came up with a third concept after extensive analysis of the height of parents and their children: the phenomenon he called *regression to the mean*. In this case, it simply means that sons of very tall fathers tended to be shorter than their fathers, and sons of very short fathers tended to be taller than their fathers. In other words, sons of fathers with unusual heights tended to be of a height that was closer to the average.

Galton used these methods in psychology, notably to study heredity, and also advanced the idea of using testing to measure psychological differences between people. As a result, he is now recognized as one of the pioneers of modern psychology.[19]

Being convinced that statistics can explain most phenomena, Galton made the first statistical study of the efficacy of prayers, by examining human longevity of various social classes and groups, including missionaries and sailors, and concluding that—alas—there was no evidence that prayers helped. In the second half of his life, he became increasingly influenced by the work of Darwin, and he came up with *eugenics*, a—by now completely discredited—idea of improving the human race through selective breeding.

When he turned 50, Galton invented an ingenious device called a *quincunx*, which showed how the bell curve naturally rose in situations involving probability,[20] and used it to illustrate his statistical ideas. It consisted of a vertical board into which identical balls were dropped from the top. When falling, the balls hit pins that were arranged in evenly spaced rows holding an increasing number of pins. The balls randomly bounced left or right and eventually fell into one of the bins at the bottom, which were of the width of the balls.

Galton went on and came up with a modification of the quincunx that allowed him to explain how, in Darwin's theory of evolution, heritable variability could occur with one species.[21] By 19th-century standards, Galton was a unique figure in the history of science. He never worked at a university, yet his research greatly influenced subsequent developments, both in statistics and psychology.

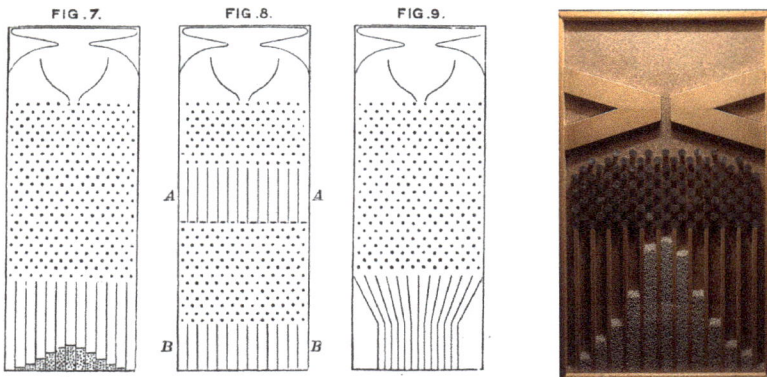

Galton's original design of a quincunx and its realization.[22]

Galton's work on statistics and views on eugenics influenced an English mathematician **Karl Pearson** (1857–1936), who applied statistics to biology to study heredity and evolution. His works, that also took place in the 20th century, led to the rise of the field of *mathematical statistics*, in which several techniques can be traced to him. Also, the term *random walk*, referring to a series of steps random in direction or length, was invented by him. Nowadays, it is occasionally used when discussing random changes in stock prices.

Another striking person in the history of statistics was an English pioneer in modern nursing and social reformer, **Florence Nightingale** (1820–1910). New nurses in many nursing schools take the pledge named after her. Nightingale was a deeply religious person but also knowledgeable in statistics: she owned a copy of Quetelet's *Social Physics* that she copiously annotated.

Nightingale served in military hospitals in Turkey during the Crimean war. She realized that high mortality was not due to the injuries in battle but to diseases and insufficient hygiene. Upon return to England, she convincingly argued for a higher standard in the field hospitals. To drive the message home, she compared the mortality statistics in field hospitals in Turkey with the ones in the vicinity of London, using innovative diagrams that were the precursors to the well-known *pie charts*.

Developments in algebra   Another important contribution from the mathematicians of the 19th century was the generalization of algebra from numbers to entities with more dimensions. Many quantities, like length, weight, or distance, can be described by just one number. But this is not the case in general. For example, to represent a store inventory, one needs to list the number of items for each product separately. Such a list of numbers is called a *vector*. The total inventory in all stores of the same company can then be readily obtained by adding the numbers in each column separately, while the total value of the inventory can be obtained by multiplying in an appropriate way the resulting vector by the one that holds the list of prices of all items.

Such considerations lead to an analysis of operations on vectors and, more generally, on *matrices*. A matrix, in the simplest two-dimensional setting, is just a table of numbers with rows and columns.

The term 'matrix' was coined by the English mathematician **James Joseph Sylvester** (1814–1897), who had a knack for inventing new terms. Perhaps this had something to do with his deep interest in the poetry that he translated into English from many languages. He also spent a total of seven years in the United States, where he twice worked as a professor of mathematics. In the 19th century, this was very uncommon. Between these two visits, he studied law and worked as an attorney. During these studies, he met another would-be lawyer, **Arthur Cayley** (1821–1895).

Cayley was an interesting character. He could not find a permanent job as a mathematician after his study of mathematics at Cambridge, in spite of having written a number of scientific papers. So he took up law studies. During his 14 years as a lawyer, he wrote some 250 mathematical papers. He treated the law as a means of earning money. Eventually, he accepted a position as a professor of mathematics at Cambridge in spite of a substantial cut in his salary.

His meeting with Sylvester was the beginning of a lasting friendship and long collaboration. Perhaps one reason was that, as in the case of Sylvester,

Cayley had many interests outside of mathematics that included literature, painting, and architecture. Inspired by Sylvester, Cayley took up a systematic study of matrices and algebraic operations on them, which laid the foundations for a whole area of mathematics now called *matrix theory*.

In the works of Sylvester and Cayley, we can once again see the power of generalization that is driving mathematics. Their idea was to treat matrices in the same way as we treat numbers and to apply the arithmetic operations of addition, subtraction, multiplication, and division to them (the last one is defined by first defining an inverse of a matrix). At first sight, this looks like yet another example of a useless occupation by mathematicians. Nothing can be farther from the truth.

To start, matrices can be used to conveniently represent data, for instance, the average income per age group (represented by rows) in various countries (represented by columns). More significantly, matrices allow us to represent linear equations in many variables in a natural way, and solving them involves then simple operations on the matrices used. Such systems of equations often appear in physics and economics, which is one of the reasons why matrices and operations on them is an indispensable tool in these disciplines.

In physics, a vector means a quantity that has a magnitude and direction, usually represented by a directed line. Nowadays, vectors and operations on them are routinely used to represent and reason about physical concepts, such as velocity, acceleration, or force.

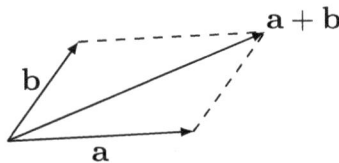

Addition of vectors.

But in the second half of the 19th century, it took a couple of decades before the appropriate algebraic formalism dealing with so-defined vectors eventually emerged. It led to a considerable simplification of the groundbreaking contributions of a Scottish physicist **James Clerk Maxwell** (1831–1879), who at the age of 30, produced his laws of electromagnetism that unified electricity and magnetism. In its final version, Maxwell had 20 equations. What is now called *Maxwell's equations* are just four equations that refer to appropriate operations on vectors.[23]

In particular, his laws allowed Maxwell to conclude that light is an electromagnetic wave, which was confirmed experimentally soon after his death. This, in turn, eventually led to the invention of radio, television, and wireless communication. Not surprisingly, Maxwell's equations are considered by many to be the greatest achievement of 19th-century physics. Maxwell was also the first scientist who applied probability and statistics in physics, to describe the distribution of speeds of molecules.

This shift from solving equations to a study of *mathematical structures*, i.e., some sets of elements (like matrices) with operations on them (like addition and multiplication), led to a new, broader understanding of what algebra actually should be concerned with. In this more general view, several natural types of mathematical structures were soon identified and systematically studied. Historically, the first was the concept of a group originally introduced by Galois and subsequently generalized and systematized by Cayley.

Developments in logic In hindsight, a spectacular development also took place in the area of logic. An Englishman, **George Boole** (1815–1864), was the son of a poor shoemaker who, at of age of 14, had to support his whole family on his own. His education ended in primary school and he was almost entirely self-taught. Yet, by the age of 20, he was already publishing original mathematical work. In 1854, as a professor of mathematics at the University of Cork in Ireland, while unaware of the unpublished works of Leibniz on a related subject, Boole published a groundbreaking book, *The Laws of Thought*. In it, he introduced a logical calculus that allowed him to systematize Aristotle's and the Stoics' approach to logic. His approach, now called *Boolean algebra* (or *Boolean logic*), connected logic with algebra by viewing the logical connectives AND, OR, and NOT alternatively as the operations on the truth values, which are identified with 0 (false) and 1 (true). His work was ignored for more than 80 years until it was found that his approach to logic could be used to describe digital circuits.[24] This, in turn, became relevant for the construction of the first computers. I shall return to this matter in the next chapter.

An important next step in the history of logic was achieved by **Gottlob Frege** (1848–1925), a German mathematician and logician with extremely rightist views. Frege embarked on an ambitious research program aimed at showing that arithmetic and parts of mathematics could be formulated in logic. In 1879, he published a book *Begriffschrift* (Concept Script), which eventually became a landmark in the history of logic. In it, he vastly generalized Boolean logic so that, in particular, statements like 'for every natural number, there is a prime number larger than it' can be formalized in it.

Unfortunately, his awkward and now defunct notation to describe logical formulas made the book difficult to typeset and publish. For example, the above statement about prime numbers would be written in his notation as

$$\vdash \underset{}{\overset{\mathfrak{a}}{\smile}} \underset{}{\overset{\mathfrak{b}}{\smile}} \begin{array}{l} prime(\mathfrak{b}) \\ greater(\mathfrak{b}, \mathfrak{a}) \end{array}$$

However, the book opened a way to represent mathematical arguments in logic. His idea was to first define the notion of a set, and subsequently, use it to introduce 0 and natural numbers, and then the operations on numbers and relations on them.

Frege continued his work in his book *Die Grundlagen der Arithmetik* (The Foundations of Arithmetic) by trying to reduce arithmetic, and subsequently, the whole of mathematics to logic, and in this way, trying to prove the consistency of mathematics. Unfortunately, in 1902, when the second volume of his book was in press, Frege received a one-page letter from **Bertrand Russell** (1872–1970), one of the most renowned philosophers of the 20th century, already mentioned a number of times. In it, Russell pointed out an important error in Frege's logic that led to a contradiction. Frege recognized the error but did not succeed in removing it. Russell's letter had a dramatic effect on Frege. He published his book with the following sad admission in the preface:[25]

> Hardly anything more unfortunate can befall a scientific writer than to have one of the foundations of his edifice shaken after the work is finished. This was the position I was placed in by a letter of Mr Bertrand Russell, just when the printing of this volume was nearing its completion.

After that, he became seriously depressed and eventually lost hope that mathematics could be founded on logic. This was not the case for Russell, whose work will be discussed further in the next chapter. Paradoxically, Russell's devastating letter and his subsequent work saved Frege from complete oblivion.

Georg Cantor   One of the most original mathematicians of the second half of the 19th century was the German (though born in Saint Petersburg) *Georg Cantor* (1845–1918). He found a way to compare and differentiate between infinite sets, which had far-reaching consequences. In particular, his approach contradicted Euclid's postulate that 'the whole is greater than the part'. A germ of this idea was present in the works of Galileo, who wrote in one of his works:[26]

> [...] we must say that there are as many squares as there are
> numbers because they are just as numerous as their roots, and all
> the numbers are roots.

But Cantor went much farther in his study. In particular, he showed by means of a novel *diagonal argument*, that the sets of reals and natural numbers are of a different infinity (called *cardinality*). By combining it with an observation that natural numbers and rational numbers are of the same cardinality, this provided an alternative and very simple proof of the existence of irrational numbers.

Cantor's work baffled his fellow mathematicians who would not accept his results on several official occasions. Their openly hostile reception eventually led to his hospitalization and a permanent depression until the end of his life. Cantor's groundbreaking achievements were recognized only after his death.

Henri Poincaré   One of Cantor's critics was the French mathematician, physicist, and philosopher **Henri Poincaré** (1854–1912). Poincaré was one of the most prominent mathematicians of his generation. He was physically frail and notoriously absent-minded, in contrast to his cousin Raymond Poincaré, who served as President of France during the First World War. Henri Poincaré was prodigiously prolific and strikingly broad in his research. During his scientific career, he wrote more than 30 books, mainly on mathematical physics and celestial mechanics. According to E.T. Bell, a historian of mathematics, Poincaré was "the last man to take practically all mathematics, both pure and applied, as his province",[27] but he also wrote popular essays and articles on the philosophy of science.

In 1887, to honor his 60th birthday, Oscar II, King of Sweden, established a prize for finding a solution to the n-body problem that generalized to an arbitrary number n, the three-body problem originally considered by Newton. In essence, the challenge was: can we derive from Newton's laws of motion, the orbits of the planets around the Sun, taking into account their mutual attraction?

This sounds like an exotic mathematical problem, but it concerns one of the most fundamental aspects of our existence. Until now, the planets have been following their orbits around the sun graciously, but how can we be sure that they will not change their course due to some accumulation of small changes or, even worse, collide?

This question, usually referred to as the *stability of the solar system*, attracted since Newton's time the attention of several prominent mathematicians, including—as already mentioned—Laplace, and was a motivation for Oscar II's prize. It was awarded to Henri Poincaré, even though he did not completely solve the problem. Moreover, he discovered that his prize-winning version contained a major error, which led him to destroy all copies and print a corrected version at his own cost.[28]

Poincaré's scientific contributions extended to the 20th century. In 1904, he put forward an important conjecture concerning spheres that was solved only 100 years later. I shall return to it briefly in the next chapter.

He was one of the first scientists who questioned the concept of the simultaneity of two events in space. In 1904, he gave a lecture in which he referred to the *relativity principle*, which states that the laws of physics have the same form for all observers. This is one of the two postulates of Einstein's special relativity theory (the other being the constancy of the speed of light for all observers) that he conceived in 1905. But Poincaré missed discovering Einstein's theory and its consequences, as he searched for additional hypotheses that were not needed.[29]

Poincaré strongly believed in intuition and famously stated: "It is by logic that we prove, but by intuition that we discover." He organized his work so that he could stimulate his subconsciousness in the periods in between; he worked two hours in the morning and two hours in the early evening.

Prime number theorem Let me conclude this account of the developments in the 19th century by mentioning an important contribution about prime numbers. Euclid's result shows that infinitely many prime numbers exist but says nothing about how frequently they occur. The following table suggests that they occur less and less often:

Percentage of prime numbers	Below
25%	100
16.8%	1,000
12.29%	10,000
9.592%	100,000
7.8498%	1,000,000

Quantifying this percentage precisely occupied the attention of several prominent 19th-century mathematicians. Gauss, at the age of 15 (!), after studying all the prime numbers up to 3,000,000, proposed an estimate related to the logarithm function. The result was eventually established independently in 1896 by two mathematicians (who both lived to an astonishingly long age)— the Frenchman **Jacques Hadamard** (1865–1963) and the Belgian **Charles**

Jean de la Vallée-Poussin (1866–1962), and who relied on complex numbers in their arguments. This result is called the *prime number theorem*.[30]

Timeline

Joseph-Louis Lagrange (1736–1813)
Pierre-Simon de Laplace (1749–1827)
Paolo Ruffini (1765–1822)
Joseph Fourier (1768–1830)
Carl Friedrich Gauss (1777–1855)
Siméon Denis Poisson (1781–1840)
Augustin-Louis Cauchy (1789–1857)
Nikolai Lobachevsky (1792–1856)
Adolphe Quetelet (1796–1874)
Niels Henrik Abel (1802–1829)
János Bolyai (1802–1860)
Peter Gustav Lejeune Dirichlet (1805–1859)
Évariste Galois (1811–1832)
James Joseph Sylvester (1814–1897)
George Boole (1815–1864)
Florence Nightingale (1820–1910)
Arthur Cayley (1821–1895)
Francis Galton (1822–1911)
Bernhard Riemann (1826–1866)
James Clerk Maxwell (1831–1879)
Georg Cantor (1845–1918)
Gottlob Frege (1848–1925)
Henri Poincaré (1854–1912)
Karl Pearson (1857–1936)
Jacques Hadamard (1865–1963)
Charles Jean de la Vallée-Poussin (1866–1962)
Bertrand Russell (1872–1970) (also discussed in the next chapter)

Notes

[1]U.C. Merzbach and C.B. Boyer, op. cit., p. 464.
[2]P. Gorroochurn, *Classic Topics on the History of Modern Mathematical Statistics*, p. 3., Wiley, 2016.
[3]U.C. Merzbach and C.B. Boyer, op. cit., p. 445.
[4]G.F. Simmons, op. cit., p. 172.
[5]G.F. Simmons, op. cit., p. 175.

[6]This theorem concerns arbitrary polynomials, so expressions such as $2x^5 - 5x^4 + 5$, or $x^6 + 3x^5 - 8x^4 + 2x^3 + 2$. It states that any polynomial of nth degree (the highest power used) that uses as coefficients complex numbers has exactly n solutions (also called zeroes) in complex numbers. In an appropriate decomposition of the polynomial, some of these solutions can be repeated.

[7]The method was actually proposed first by the French mathematician A.-M. Legendre, with whom Gauss got into a bitter dispute about its priority.

[8]Each IBAN is a string of up to 34 signs that are either letters or digits. It is converted according to a fixed procedure to a large number N. Then, one checks whether $N \bmod 97 = 1$. If it is the case, the IBAN is validated.

[9]M. Livio, op. cit., pp. 86–89.

[10]For an account of Abel's life, see M. Livio, op. cit., pp. 90–111.

[11]The English translation of this letter is available from the website of the Indian Academy of Sciences at `https://www.ias.ac.in/article/fulltext/reso/004/10/0093-0100`.

[12]For an account of Galois's life, see M. Livio, op. cit., pp. 112–157.

[13]A group is simply a set of elements with a distinguished element, called the *identity* and a single operation on them that satisfies a couple of intuitive properties. An example of a group is the set of clockwise rotations of an object by the multiple of 10 degrees. The rotation by 0 degrees is the identity. The rotations can be combined. The crucial property is that any rotation can be undone (reversed), by combining it with another one. For instance, to undo the rotation by 20 degrees, one needs to rotate the object further by 340 degrees.

In contrast, the clockwise rotations by a multiple of 7 degrees do not form a group. Indeed, the rotation by 7 degrees cannot be undone, as 353 is not a multiple of 7.

[14]B. Belhoste, *Augustin-Louis Cauchy: A Biography*, Springer-Verlag, p. 112, 1991.

[15]V.J. Katz, op. cit., p. 783.

[16]It was found recently that this principle was mentioned two centuries earlier by a French Jesuit J. Leurechon, see B. Rittaud and A. Heeffer, The Pigeonhole principle, two centuries before Dirichlet, *The Mathematical Intelligencer*, 36(2), pp. 27–-29, 2014.

[17]D.J. Boorstin, op. cit., pp. 672–675.

[18]See for example, H. Tijms, op. cit., pp. 105–106.

[19]R.F. Fancher, *Pioneers in Psychology*, W.W. Norton & Company, 3rd edition, pp. 216–245, 1996.

[20]More precisely, it illustrates the result of de Moivre that Bernoulli trials converge to the normal distribution.

[21]S.M. Stigler, *The History of Statistics: The Measurement of Uncertainty before 1900*, Belknap Press, pp. 276–277, 1986, and S.M. Stigler, *The Seven Pillars of Statistical Wisdom*, Harvard University Press, pp. 111–132, 2016.

[22]Courtesy: Public Domain, Wikipedia Commons.

[23]This simplification was achieved in 1884 by the English mathematician and physicist O. Heaviside.

[24]D. Mac Hales, *George Boole*, in: T. Gowers, I. Leader, and J .Barrow-Green (eds.), *The Princeton Companion to Mathematics*, Princeton University Press, pp. 769–770, 2008.

[25]G. Frege, *The Foundations of Arithmetic*, Volume II, Routledge, 2007 (originally published in 1903).

[26]Galileo Galilei, *Dialogues Concerning Two New Sciences*, Cosimo Classics, 2010 (originally published in 1638).

[27]E.T. Bell, *Men of Mathematics*, Simon and Schuster, p. 527, 1965.

[28]I. Stewart, *In Pursuit of the Unknown: 17 Equations That Changed the World*, Basic Books, p. 64, 2012.

[29]See A. Pais, *Subtle Is the Lord: The Science and the Life of Albert Einstein*, Oxford

University Press, pp. 128 and 168, 1982.

[30] Hadamard, apart from being a prominent mathematician, also initiated a study of the psychology of mathematical thinking, a topic that interested Poincaré as well. After having interviewed several physicists, Hadamard concluded that the creative process consists of the following four stages: preparation, incubation, illumination, and verification.

Chapter 8

The 20th and 21st Centuries

The 20th century is sometimes referred to as the 'golden age of mathematics'. It is characterized by a veritable explosion and fragmentation of the subject. It has been estimated that in 1870, about 840 papers in mathematics were published, while towards the end of the 20th century, some 50,000 papers were published annually.[1] As a result, it is difficult, if not impossible, to provide a comprehensive account of all developments, especially for a layman.

In the works of 19th-century mathematicians, we already saw that such concepts like an irrational number, function, or a limit can be defined precisely. The 20th-century mathematicians went much further. Their works not only clarified other concepts, for example, distance, measure, dimension, set, a computable function, or randomness, to name a few, but also led to the development of notions and theories that put such disciplines as logic, algebra, geometry, analysis (which, as already mentioned, includes calculus), or probability on firm grounds. Through vast generalizations of calculus, algebra, and geometry, new, abstract, areas of mathematics were developed. Most of these advances cannot be explained without discussing mathematical concepts. What follows, therefore, is an exposition of some easier-to-explain developments with an unavoidably incomplete reference to the dramatis personae responsible for the spectacular growth of the subject.

Morris Kline, a previously mentioned historian of mathematics, wrote that "the search for generality and unification is one of the distinctive features of twentieth century mathematics [. . .]".[2] Another distinctive feature is a study of its foundations, so investigation of the basic concepts using which mathematics can be defined.

David Hilbert   A leading figure behind these developments was a German mathematician **David Hilbert** (1862–1943), one of the most prominent mathematicians of the first half of the 20th century. He was born in Königsberg, where he attended the same school as Immanuel Kant 140 years earlier. Eventually, he moved to Göttingen, which was a leading center of mathematics until the beginning of the Second World War.

His initial contributions were published at the end of the 19th century. Among them are fundamental results that linked algebra and geometry in a more profound way than analytic geometry does. This led to the rise of the field of *algebraic geometry*, to which many prominent 20th-century mathematicians contributed. It is concerned with curves or surfaces that can be described as solutions of algebraic equations. An early example of such work is the Apollonius of Perga study of conic sections because each of them can be described by an equation of the second degree involving two variables. But Hilbert went much further by investigating properties described as common zeroes of polynomials of several variables.

A down-to-earth example of an algebraic geometry problem is the *piano mover's problem*, which is concerned with finding a continuous motion of an object ('a piano') from an origin ('a storage room') to the destination ('the living room') that avoids collisions with the walls. Nowadays, such problems naturally arise in robotics, and their solutions call for a combination of techniques drawn from algebraic geometry and computer science.

Hilbert's major work was the 1899 book *Grundlagen der Geometrie* (Foundations of Geometry), which provided a way to reduce geometry to the theory of real numbers. Hilbert achieved this by introducing three primitive notions: point, line, and plane, and three relations linking them, for instance, 'between' involving three points. This allowed him to express Euclid's axioms by 20 axioms formulated in a formalism amenable to a rigorous study. In the process, Hilbert also discovered and filled some gaps in the arguments given in Euclid's *Elements*. This is viewed as the origin of *formalism*, an approach to mathematics characterized by its reliance on the deductive method.

In 1900 (admittedly still in the 19th century), during the International Congress of Mathematicians, Hilbert announced a list of 23 open mathematical problems. This list became highly influential and stood at the outset of many subsequent developments. The still unsolved Riemann hypothesis, already mentioned in the previous chapter, figures on this list.[3]

One of the areas to which Hilbert contributed in the 20th century was physics. In 1915, Einstein proposed the general relativity theory that extended his 1905 special relativity theory by incorporating gravity into it. After he explained his progress in developing the theory in a series of lectures that he gave at the University of Göttingen in the summer of 1915, Hilbert enthusiastically endorsed the underlying ideas. As a result, he started to work on the subject himself and was in fact on the track to discover the final theory first.

In contrast to critics of Cantor, Hilbert was highly impressed by his contributions. He once said: "No one shall expel us from the paradise which Cantor created for us."[4]

Soon after Cantor's death, Hilbert found a nice illustration of Cantor's definition of an infinite set, by proposing what is now called *Hilbert's hotel*. In this hotel, there are infinitely many rooms. Suppose all of them are occupied. This does not prevent the receptionist from accommodating a newly arrived guest. He simply asks the guest from room 1 to move to room 2, the guest from room 2 to move to room 3, etc. This way, all guests can still keep a room and, moreover, the new guest can move to room 1. This trick, of course, does not work in the customary hotels that have 'only' finitely many rooms, but it illustrates the defining property of an infinite set (here, the one consisting of 1, 2, 3, . . .); namely, that it has as many elements as its subset (here, the one consisting of 2, 3, 4, . . .).

Foundational crisis of mathematics The letter that Russell sent in 1902 to Frege showed that a naive understanding of the notion of a set led to a contradiction. The resulting complication is now called *Russell's paradox*. It is discussed in Appendix 31.[5] Since the notion of set is fundamental in all mathematics, this paradox led to the so-called *foundational crisis of mathematics*.

It motivated Russell to write the highly influential *Principia Mathematica*, jointly with his mentor and close friend **Alfred North Whitehead** (1861–1947), which attempted to construct nothing less than a logical system from which all mathematical truth could be derived. This view is called *logicism*, and Russell is considered to be the main proponent of it. The book took the authors ten years to write. The outcome, in the manuscript of 4,500 pages and published in three volumes from 1910 to 1913, was hardly readable. Famously, the proof of the (in the authors' words) 'occasionally useful proposition' that $1 + 1 = 2$ took 10 lines of calculations.

Years later, Russell admitted:[6]

> I used to know of only six people who had read the later parts of the book. Three of these were Poles, subsequently (I believe) liquidated by Hitler. The other three were Texans, subsequently successfully assimilated.

Notwithstanding, the message of this treatise that mathematics could be

*54·43. $\vdash :. \alpha, \beta \in 1 . \supset : \alpha \cap \beta = \Lambda . \equiv . \alpha \cup \beta \in 2$

Dem.

$\vdash . *54\cdot26 . \supset \vdash :. \alpha = \iota^{\iota}x . \beta = \iota^{\iota}y . \supset : \alpha \cup \beta \in 2 . \equiv . x \neq y .$

[*51·231] $\equiv . \iota^{\iota}x \cap \iota^{\iota}y = \Lambda .$

[*13·12] $\equiv . \alpha \cap \beta = \Lambda$ (1)

$\vdash . (1) . *11\cdot11\cdot35 . \supset$

$\vdash :. (\exists x, y) . \alpha = \iota^{\iota}x . \beta = \iota^{\iota}y . \supset : \alpha \cup \beta \in 2 . \equiv . \alpha \cap \beta = \Lambda$ (2)

$\vdash . (2) . *11\cdot54 . *52\cdot1 . \supset \vdash . \text{Prop}$

From this proposition it will follow, when arithmetical addition has been defined, that $1 + 1 = 2$.

The proof that $1 + 1 = 2$ from *Principia Mathematica*.[7]

reduced to logic had a huge impact. It inspired Hilbert to propose in 1921 a research program, subsequently called *Hilbert's program*, that was the culmination of his continued interest in the foundations of mathematics. Its purpose was to solve the foundational crisis in mathematics by reducing the consistency of mathematics to that of arithmetic.

Hilbert's work led to the creation of *predicate logic* (sometimes called *first-order logic*) that extends the propositional logic with the quantifiers 'for all' and 'for some'. This has had important consequences for mathematics since various mathematical statements can be expressed in predicate logic.

In a book with a coworker, Hilbert also asked whether there is a procedure that allows us to determine whether a sentence in the predicate logic is true. If there is, then, in principle, the work of many mathematicians would be superfluous: if a theorem could be expressed using predicate logic, one could instead just apply such a procedure. This problem, called in German, *Entscheidungsproblem* (decidability problem), was answered in the negative in 1936, independently by Alonzo Church and Alan Turing. In Appendix 25, I provide a minimal introduction to predicate logic.

Kurt Gödel Hilbert's program was shattered by a young Austrian, **Kurt Gödel** (1906–1978), who became the most prominent logician of the 20th century. In 1931, at the age of 25, he published a groundbreaking paper in which he showed that the theory of arithmetic (which involves axioms about natural numbers; for instance, $x + y = y + x$ or for no x, we have $x + 1 = 0$) is too weak to prove all true statements about natural numbers. Further, he showed that the consistency of arithmetic can be formalized in arithmetic but cannot be proved in it (unless arithmetic is inconsistent).

In other words: if the consistency of arithmetic can be proved in it, then it is actually inconsistent. This is a result with far-reaching philosophical consequences about mathematics because it applies to all proof systems that include arithmetic.

In his proof, Gödel cleverly employed Cantor's diagonal argument to a version of the Liar paradox credited to Eubulides, a Greek philosopher from the 4th century BCE. (Someone says: 'This statement is false.' The question is: Is this statement true or false?) Gödel's proof employed a novel encoding of arithmetic formulas and their formal proofs by numbers that relied on the Chinese remainder theorem, first mentioned in Chapter 3. Gödel's results and their original proofs had a tremendous impact on the subsequent developments in logic and philosophy of the 20th century.

Gödel was one of the many scientists who left Europe for the United States after the rise of Nazism, though he hesitated until the last moment. Eventually, together with his wife, he left Austria only in 1940 and reached the United States, by a roundabout way by taking the Trans-Siberian Railway to Vladivostok, followed by ships to Yokohama, and from there to San Francisco. At Princeton, he worked at the same institute as Einstein, who was one of the very few people he kept in touch with. Tragically, Gödel developed an obsessive fear of being poisoned, which led to his death from starvation—a tragic paradox akin to his own theorem.

Through discussions with Einstein, Gödel became interested in general relativity theory and proposed an intriguing solution in which one could travel backward in time. Interestingly, in the early seventies, Gödel also provided a logical proof of God's existence, though—not surprisingly—many philosophers question the axioms used in it.[8]

At Princeton, Gödel also established important results in set theory, an area of mathematics initiated by Georg Cantor. Deep work of a young American mathematician **Paul Cohen** (1934–2007) built upon them, which led him to a solution of the first problem on Hilbert's list. The problem, called the *continuum hypothesis*, is concerned with the nature of the infinity of the set of real numbers. It was originally posed by Cantor, who tried in vain to prove it for several years.

Alfred Tarski | Another leading logician of the 20th century was **Alfred Tarski** (1901–1983), a Pole who emigrated to the United States on the eve of the Second World War. His definition of *truth* in formal languages was adopted by philosophers, logicians, and linguists, and his result on the undefinability of truth in arithmetic led to a better understanding of the original Gödel result. Tarski also suggested a novel way to resolve the Liar paradox by distinguishing between a language (in which statements are constructed) and a meta-language (in which statements about the original statements are made).

Perhaps the biggest achievement of Tarski was the discovery of an algorithm that allows one to decide the truth of predicate logic statements about real numbers. This, combined with Descartes' discovery that geometrical problems can be expressed using statements about real numbers, allows us, in principle, to determine the truth of many geometric problems, in particular, all problems of Greek geometry. As shown later, this algorithm is too inefficient to be used in practice, though it gave rise to *computer algebra*, an area at the intersection of computer science and mathematics, the aim of which is to develop algorithms to manipulate mathematical expressions.

In contrast to the solemn and ascetic Gödel, Tarski was a colorful socialite. The expression 'Sex, Drugs, and Rock and Roll' applies very well to him, provided one replaces 'Rock and Roll' with 'Alcohol'.[9]

Alan Turing   The already mentioned ***Alan Turing*** (1912–1954) was an English logician and mathematician. In 1936, inspired by the results of Gödel, he formulated a simple paper model of computing, now called a *Turing machine*. Its design vastly generalized Pascal's Pascaline calculating machine and Leibniz's calculator since it allowed one to execute an arbitrary sequence of instructions and not only the arithmetic operations. Ten years later, it inspired the actual building of one of the first computers.[10] The resulting notion of computation of such a machine also stood at the cradle of *theoretical computer science*, which, among others, focuses on the design of algorithms and the analysis of their efficiency.

Turing also proposed a test, now called *Turing's test*, which is supposed to distinguish between a human being and a computer program. The idea is to determine 'who is who' from a transcript of a natural language interaction between a human being and a computer program. If a computer program can fool the judges, then it passes the test. This idea made Turing one of the pioneers of *artificial intelligence*, a broad discipline that attempts to capture the cognitive aspects of human behavior (such as reasoning, planning, learning, decision making, problem solving, vision, etc.), aiming at replacing human actions by computers.

In 1945, in recognition of his work on deciphering codes used by the Germans during the Second World War, Turing received the Order of the British Empire. A few years later, his life took a dramatic turn. In 1952, he was arrested for violating British homosexuality laws and was subjected to estrogen injections for a year. Two years later, he was found dead after having eaten an apple filled, probably by himself, with cyanide.

A portrait of Alan Turing on the £50 banknote.[11]

To recognize Turing's contributions to the foundations of computing, the main association in computer science, the Association for Computing Machinery (ACM), established an annual prize called the ACM A.M. Turing Award in 1966. It is considered a computer science counterpart of the Nobel Prize. In 2009, British Prime Minister Gordon Brown officially apologized for Turing's treatment and a couple of years later Queen Elizabeth II officially pardoned him. In 2021, the Bank of England issued a banknote with the portrait of Alan Turing on it.[12]

Probability theory revisited So far, I have only discussed the developments in logic. Another area where important advances were made was probability theory. Properly defining probability has bothered mathematicians ever since the origins of the subject. In particular, it was not clear how to define elementary concepts like 'probability of an event', 'equally probable' or 'more likely'. This led Bertrand Russell to quip in 1929: "Probability is the most important concept in modern science, especially as nobody has the slightest notion what it means."[13]

The challenge was picked up four years later by the Russian ***Andrey Nikolaevich Kolmogorov*** (1903–1987), one of the most prominent mathematicians of the 20th century, whose exceptionally broad interests resulted in numerous publications in several areas of mathematics, as well as in biology, mechanics, philosophy, logic, the history of mathematics, and a quantitative study of Russian poetry. In response to one of Hilbert's 23 problems, Kolmogorov came up with a simple and convincing axiomatic definition of probability, which is now a standard approach. It is explained in Appendix 26.

The formalization of the concept of probability, no matter how convincing, can still easily clash with our intuition. I have already mentioned the Monty Hall problem in Chapter 6. Another troubling example is the following *Steinhaus–Trybuła* paradox proposed in 1959 by two Polish mathematicians.[14] Imagine that three bus lines serve the same route, but the buses arrive somewhat unpredictably. It is then possible that we are not able to determine which bus line will most likely arrive first. This paradox bears some similarity with the Condorcet voting paradox from Chapter 6. It is explained in Appendix 30.

A different, more practical approach to probability and statistics was put forward by the Englishman **Ronald Fisher** (1890–1962). In his views and approach to statistics, he was highly influenced by the work and opinions of Francis Galton, whom I discussed in the previous chapter. In particular, Fisher embraced eugenics, which led him to an extensive research in genetics. In 1930, he published a book in which he outlined a model of evolution, now called *Fisher's principle*. It explains why for species that produce offspring through sexual reproduction, the sex ratio between males and females is approximately 1:1.

Five years later, he published an influential book titled *The Design of Experiments*, in which he discussed his by now famous *lady tasting tea* experiment. The purpose of the experiment was to verify the claim of a lady who declared at a summer tea party that she was able to tell whether the tea or the milk was added first to a cup. Fisher arranged for eight cups of tea, four of which had tea poured into milk and the other four had milk poured into tea. He then calculated the probability of a given outcome. The lady in question got all eight cups correct. The chance that this happens is 1 in 70.

This experiment was highly influential in the subsequent applications of statistics, which started to be based on carefully designed experiments. Since then, statistics has been used in several fields of science, in particular, psychology, medicine, economics, sociology, astronomy, geology, and criminology, and also in stock market analysis and weather prediction.

Claude Shannon and Norbert Wiener   Towards the end of the 19th century, probability started to be applied in physics to study systems whose exact states are unknown. This led to the rise of *statistical mechanics*, which combines physics with probability theory. One of the introduced concepts was *entropy*, a measure of how probable a given physical state is. The more likely it is, the higher its entropy.

This notion inspired an American electrical engineer and mathematician, **Claude Shannon** (1916–2001), to found *information theory* in 1948, which turned out to be crucial for our understanding of the role of information and the way we can quantify its contents and process of its transmission. Shannon embraced the concept of a *bit* (coined by his colleague as an abbreviation for 'binary digit'), which he defined as the smallest unit of information, and came up with the concepts of *Shannon entropy*, which measures the amount of information contained in a message in bits, and *channel capacity*, which captures the maximum rate at which information can be faithfully passed through a communication channel. Nowadays, these concepts allow us to discuss various aspects of the World Wide Web in precise terms.

Shannon was one of the most original scientists of his times, exceptionally broad in his scientific interests. At the age of 21, he came up with the revolutionary idea, mentioned on page 104, of using Boolean algebra to describe digital circuits. This was one of the crucial insights that led to the construction of the first computers some ten years later.

Shannon was known for several unusual interests and hobbies, including riding a unicycle while juggling. In 1949, he wrote an influential article suggesting how to program a computer to play chess. In the late 1950s, he became interested in the stock market, to which he applied his insights on how to profit from fluctuations in stocks, a scheme now called *Shannon's Daemon*.

In 1960, Shannon was approached by another mathematician who convinced him to construct the first wearable computer, the only purpose of which was to beat roulette. They built it the size of a cigarette pack and eventually tested it in Las Vegas, but minor hardware problems prevented them from serious betting. Shannon also designed and constructed an amusing and completely useless machine, the only purpose of which was to switch itself off after it was switched on.[15]

Shannon was very reluctant to publish his findings and even discuss them with his colleagues. Yet his work and ideas quickly became highly influential. Marshall McLuhan, a Canadian philosopher known for his ideas on media, which were widely discussed in the 1960s, coined the familiar term 'the information age' after having been influenced by Shannon.[16]

Another noteworthy American of this period was the child prodigy **Norbert Wiener** (1894–1964), who attained his PhD from Harvard University at the age of 18. Subsequently, he spent a couple of years in Europe, where he studied under Russell in Cambridge and Hilbert in Göttingen.

Wiener made several important contributions to mathematics—in particular to harmonic analysis—and to mathematical physics, but he is most known for introducing *cybernetics* (a term coined by him). Cybernetics is concerned with the general study of control and communication in systems construed in the broadest sense, including those occurring in physics, engineering, biology, and psychology.

In 1950, he presented his views in an ambitious and broad book, *The Human Use Of Human Beings: Cybernetics And Society*. Some of these views were ahead of his time. For example, he argued for the advantages of automation in times when the number of computers could be counted on the fingers of one hand. However, he was also aware of the resulting ethical issues. In particular, he warned early on about the permanent damage we may cause to our environment if we overuse natural resources. The following quote from the book is particularly striking:

> The more we get out of the world, the less we leave, and in the long run we shall have to pay our debts at a time that may be very inconvenient for our own survival.

After the initial hype, cybernetics gradually fell out of fashion, though it enriched us with the now universally used notion of *feedback*. Wiener was exceptionally broad in his mathematical interests but was also known for his self-praising, chaotic writing style and for being a notoriously bad lecturer.[17]

Dynamical systems In nature, various phenomena, such as the weather or planetary or biological systems, obey certain rules. These rules describe, by means of equations, the way the system evolves in time. Poincaré's influential work on the three-body problem in relation to the Oscar II prize was an early example of research on such systems. In the second half of the 20th century, the name *dynamical systems* was coined for these systems, and a number of important contributions was made to this subject. Mathematicians working on dynamical systems could use computers to perform calculations and simulations and subsequently apply them to analyze outcomes and draw theoretical conclusions.

An early example of a dynamical system is the *Lotka–Volterra* model proposed independently in the 1920s by two eponymous mathematicians. It describes the evolution of a biological system involving two species, predators and prey, for example, foxes and rabbits, and allows one to explain an increase in predators or prey in specific situations. It is one of the earliest models in *mathematical ecology*. Since then, it has been used in other disciplines, including epidemics and economics.

As already noticed by Poincaré, some dynamical systems turn out to be highly sensitive to initial conditions, which can lead to highly unpredictable behavior. Such dynamical systems are the subject of *chaos theory*. One of its pioneers was an American mathematician, **Edward Norton Lorenz** (1917–2008). His famous 1963 lecture titled *Does the flap of a butterfly's wings in Brazil set off a tornado in Texas?*, now commonly referred to as the *butterfly effect*, refers to the phenomenon that small causes may result in large effects. Such effects account for difficulties in weather prediction, the main area of Lorenz's interest. Other applications of chaos theory include stock market analysis, study of the population dynamics, and fluid dynamics.

Another example of dynamical systems is *fractals*, an invention of **Benoit Mandelbrot** (1924–2010), a Polish-French-American mathematician. Intuitively, these are geometric shapes that look similar, independently of scale. Interestingly, there are many examples of fractals that are outcomes of iterations of very simple functions defined on complex numbers, or more precisely, on the plane representing them, a representation mentioned on page 55.

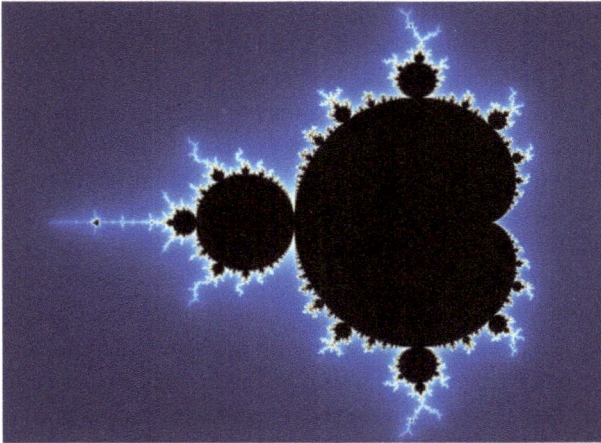

The Mandelbrot set: an example of a fractal.[18]

Mandelbrot introduced fractals in an influential 1967 paper, *How long is the coast of Britain*, which formalized the observation that the length of a coastline depends on the adopted scale. (Think of determining the length of a coast by following its twists and turns first on the meter scale, then on the centimeter scale, etc.) This brought him the idea of associating the *fractal dimension* with such objects, not necessarily a natural number, which quantifies their self-similarity. In 1975, Mandelbrot followed up with an influential and beautifully illustrated book, *The Fractal Geometry of Nature*, which was accessible to the general public.

Since then, fractals have been used to model various phenomena in a number of areas, including archaeology, geography, and medicine, and also led to the rise of *fractal art*, concerned with stunningly beautiful pictures produced by means of computer programs that generate fractals. Another application area was proposed by Mandelbrot himself who, in 2006, coauthored a book, *The (Mis)behavior of Markets: A Fractal View of Financial Turbulence*, that criticized modern financial theories for not taking into account the chaotic character of financial markets and argued instead for the use of fractals to minimize risk. I shall return to this subject in a moment.

Rise of cooperation   Throughout the centuries, mathematicians worked mostly alone. This changed after the First World War and led to the creation of some schools of thought in mathematics and logic that were characterized by commonly pursued topics and geographic proximity. Examples are the Polish school of mathematics and the Vienna circle, to which Gödel belonged and which gathered the most prominent Viennese mathematicians, philosophers, and logicians.

The 20th century also saw some impressive collaborative efforts. One of them was initiated by a group of leading French mathematicians who, starting in 1935, published several books under the pseudonym **Nicolas Bourbaki** and whose aim was to redefine the whole of mathematics. Their approach is in some ways related to the *structuralist movement*, which was highly influential in linguistics around that time.[19] In spite of the large impact of this massive work, its reception was mixed because of the highly formal and abstract style (with no pictures). For example, **Vladimir Igorevich Arnold** (1937–2010), a prominent Russian mathematician, who at the age of 19 solved one of Hilbert's 23 problems and established an important result about the n-body problem with his mentor Kolmogorov, bluntly stated:[20]

> It is awful to think what kind of pressure the Bourbakists put on (evidently nonsilly) students to reduce them to formal machines! This kind of formalized education is completely useless for any practical problem and even dangerous, leading to Chernobyl-type events.

The most spectacular example of a collaborative project is the *classification of finite simple groups*, a project that took about 100 years to complete after its formulation. (Recall that the notion of a group was introduced in Chapter 7 in the context of Galois' work on solving equations of the fifth degree.) The project involved over 100 mathematicians, resulted in more than 500 papers that totaled more than 15,000 pages, and led to one single final result in 1980.

One of the cornerstones of this classification was a paper with a proof running over 255 pages.[21]

Developments in calculus   In 1933, a prominent Russian mathematician, Nikolai Nikolaevich Luzin, wrote a letter to a colleague mathematician in which he stated:

> I cannot but see a stark contradiction between the intuitively clear fundamental formulas of the integral calculus and the incomparably artificial and complex work of the 'justification' and their 'proofs'. One must be quite stupid not to see this at once, and quite careless if, after having seen this, one can get used to this artificial, logical atmosphere, and can later on forget this stark contradiction.[22]

Luzin's comments may explain why the work on the foundations of calculus did not stop with Riemann's approach. Towards the end of the 19th century, mathematicians working in analysis became increasingly aware of some inconveniences of his definition of the integral. In particular, some examples of natural functions were found for which his definition of integral was not applicable, and it was not clear for which functions the fundamental theorem of calculus mentioned in Chapter 5 was applicable.

To address some of these shortcomings, a French mathematician **Henri Léon Lebesgue** (1875–1941) proposed in 1902 in his influential PhD thesis a radically alternative approach to integration based on the concept of a *measure*, which sought to formalize the idea of the size of a set of real numbers. It led to a more general definition of integration than that of Riemann and revolutionized the understanding of analysis.

The details of his approach are beyond the scope of this book. Lebesgue offered the following intuition to a colleague:

> I have to pay a certain sum; I look through my pockets, and there I find coins and currency notes of various values. I give them to my creditor in the order in which I find them until I have reached the total amount of my debt. That is the Riemann integral. But I proceed otherwise. After taking all the money out of my pockets, I place all of the notes of the same value together, and I do the same with the coins, and I make my payment by handing over in sequence all the money of a given value. That is my integral.[23]

However, Lebesgue's approach did not supersede the one by Riemann, and even during mathematics studies, one usually begins with Riemann's definition of integration. An American mathematician Richard Hamming, one of the first recipients of the Turing Award, once quipped: "Does anyone believe that the difference between the Lebesgue and Riemann integrals can have physical significance, and that whether, say, an airplane would or would not fly could depend on this difference? If such were claimed, I should not care to fly in that plane."[24]

Over the next 50 years, a number of mathematicians came up with modifications of Riemann's approach, which led to a definition of an integral that in some situations, was even more general than the Lebesgue integral. In 1997, six book authors distributed an open letter to the authors of calculus textbooks, where they argued that the Riemann integral should be accompanied by a slightly more general definition that would allow one to formulate the fundamental results of calculus in a less restricted way.[25] Judging by the most popular standard calculus books on the market, their letter did not have (yet?) the desired effect.

After the works of 19th-century mathematicians that resulted in rigorous foundations of calculus, the original informal idea of Newton and Leibniz of using 'infinitesimals' seemed long forgotten. So it came as a surprise that it was possible to provide an alternative, rigorous approach to calculus based on infinitesimals after all. It was suggested in the early 1960s by an American mathematician **Abraham Robinson** (1918–1974), and worked out in a number of textbooks. In this approach, called *nonstandard analysis*, a model of real numbers is constructed in which infinitesimals exist.[26] Reception of nonstandard analysis differed greatly among mathematicians. It is fair to say that it did not lead to a 'dethroning' of Riemann's or Lebesgue's approach.

Game theory The 20th century also saw the rise of *mathematical economics*, the aim of which is to model economic interactions using mathematics. Examples of such phenomena are competition between companies that target the same group of customers, a market in which buyers and sellers buy and sell goods at agreed prices, bargaining, or the problem of a fair division of costs among all users of a common facility. In each case, mathematical techniques allow one to describe, analyze, and solve the underlying problem.

To this purpose, economists use the whole arsenal of mathematical techniques. One of the main tools is *game theory*, a vast subject that comprises chess, parlor games, and also games that capture the economic interactions mentioned above.

What precisely is a 'game', and hence, game theory, is not easy to define. Ludwig Wittgenstein summarized the difficulty in 1953 in his *Philosophical Investigations* as follows:

> Don't say: 'There must be something common, or they would not be called "games"'—but look and see whether there is anything common to all. For if you look at them, you will not see something that is common to all, but similarities, relationships, and a whole series of them at that. To repeat: don't think, but look!

In the economic context, a game is viewed as an interaction between players and can be either cooperative or non-cooperative (also called strategic). In general, players in cooperative games aim to find a solution satisfactory to all, while in strategic games, each player aims to maximize his own profit. An example of a cooperative game is the bankruptcy problem, concerned with a fair division of the estate among the creditors, when the total of their claims exceeds the value of the estate.

The most famous example of a strategic game is the *prisoner's dilemma* game identified in 1950. Perhaps the simplest way to describe it is by considering an arms race between two countries equal in strength. Each country has two options: to invest in arms or not. If a country fails to make such investments, the other one will, and in this way, profit from the military advantage (for example, by exercising pressure in bilateral economic negotiations). As a result, it is better for both countries to invest in arms. But this yields a worse outcome for each country than the one obtained by means of bilateral disarmament.

This dilemma cannot be solved by any prior discussion of its consequences. It has led to thousands of papers in many fields, including psychology and sociology. A more precise account of the prisoner's dilemma is given in Appendix 32.

Other examples of strategic games are ones in which players repeatedly take turns to make moves, for instance, chess, checkers, or Go. In such games, the crucial question is to determine a winning strategy or a strategy that guarantees at least a draw. Such games are too complex to be analyzed mathematically, but nowadays, their analysis is possible by means of computer programs.[27]

Game theory has also been successfully used in other sciences, notably in biology to explain evolution, in sociology to explain the rise of racial segregation, and in law to explain rational decisions.

John von Neumann and John Forbes Nash, Jr.

Mathematically, the main results in strategic games are due to **John von Neumann** (1903–1957), who studied two-player zero-sum games (and coined their name), and **John Forbes Nash, Jr.**

(1928–2015), who at the age of 21, extended von Neumann's approach to arbitrary strategic games by introducing a concept now called the *Nash equilibrium*. Both of them were striking figures.

Von Neumann was a prominent Hungarian-American mathematician, a child prodigy famous for his incredible speed in computing and fast understanding of involved mathematical problems. Throughout his life, he had exceptionally broad interests that led him to fundamental contributions in several areas of pure and applied mathematics, starting with logic (in particular, upon hearing Gödel's brief announcement of his first result, von Neumann independently discovered the second one) and involving various applications of mathematics in physics (notably in quantum mechanics), economics (where he proposed a model of stable economic growth and co-authored a groundbreaking book on game theory), and computer science (where he suggested the universally adopted computer architecture, now called the von Neumann architecture, and came up with a model of a self-replicating automaton).

In 1943, von Neumann joined the Los Alamos Manhattan project, the aim of which was to build the American atomic bomb, and contributed to it notably by developing mathematical models of implosion. After the war, he got convinced about the importance of computers and got involved in developing one of the first general-purpose computers, ENIAC (Electronic Numerical Integrator and Computer) that was completed at the end of 1945. It was a 'monster' that weighed about 30 tons. It could perform 2,000 multiplications in one second. (For comparison, typical laptops nowadays can execute a couple of billion (10^9) instructions in one second.) Von Neumann quickly realized its potential and, when questioned about his interest in developing more powerful bombs, prophetically replied: "I am thinking about something much more important than bombs; I am thinking about computers."

Von Neumann's work on self-replicating automata influenced research in neurology and had an impact on the budding field of artificial intelligence by leading to research on *artificial life*. In his unfinished and posthumously published book *The Computer and the Brain*, he suggested, probably as the first scientist to do so, analogies between the human brain and a computer.

After the war, von Neumann became a consultant and member of a number of important US agencies, including the Atomic Energy Commission and the United States Air Force. He died prematurely of cancer, probably due to exposure to radiation during the first atomic blast that took place in New Mexico and that he attended. Shortly before his death, much to the surprise of his friends and colleagues, he converted to Catholicism by embracing Pascal's wager.[28]

John Nash, before succumbing to paranoid schizophrenia at the age of 30, was considered one of the most talented mathematicians of his generation, with brilliant contributions to geometry and to partial differential equations.[29] Several years later, he recovered from his illness and, in 1994, shared the Nobel Memorial Prize in Economic Sciences for his contributions to game theory. This remarkable story was immortalized in the book, *Beautiful Mind* (by Sylvia Nasar), which was subsequently turned into a successful albeit somewhat romanticized film. In yet another twist of fate, Nash's life had a tragic end. In 2015, he received the prestigious Abel Prize in Norway for his mathematical achievements. Upon returning to the US, on the way back from the airport, he and his wife were killed in a car accident caused by their taxi driver.

Other developments in mathematical economics

Von Neumann and Nash are just two researchers who contributed to the 'mathematization' of economics. Other mathematicians found new uses of mathematics in economics. Let me mention three examples here.

Voting procedures have a long history starting with the Athenian democracy from the 6th century BCE, when decisions were taken during assembly meetings, of which members were adult male citizens who were not slaves. The voting was by simple majority. Over the centuries, many other voting procedures were developed. I already mentioned the methods advocated by Ramon Llull, Marquis de Condorcet, and Jean-Charles de Borda. However, each of these methods had some drawbacks.

In 1950, **Kenneth Arrow** (1921–2017) established in his PhD thesis a surprising result: it is impossible to construct a voting procedure that satisfies some simple and desired properties. His result is occasionally summarized misleadingly as a statement that 'dictatorship is unavoidable'. This is, of course, nonsense. Arrow was concerned with the problem of aggregating individual preferences into a collective preference, so in his setting—unlike in the customary elections—the voters ranked all candidates. Further, one of his assumptions, called 'independence of irrelevant alternatives', is occasionally considered debatable. This innovative work initiated the field of *social choice theory* concerned with the analysis of voting, ranking procedures, and elections.

Arrow, jointly with **Gérard Debreu** (1921–2004), also formalized and proved the 'invisible hand' principle in a market economy, suggested by the 18th-century economist Adam Smith. This work is part of the field of *welfare economics*. For their contributions to mathematical economics, each of them received the Nobel Memorial Prize in Economics.

It has to be added that several influential mathematical results in the economics and social sciences have surprisingly short proofs. For instance, the
fundamental Nash theorem about the existence of equilibria in strategic games
is a one-page long corollary to a mathematical result called Brouwer's fixed-
point theorem, mentioned on page 135, and the shortest proof of Arrow's
theorem about voting procedures now fits one page, as well.

There are also applications of mathematics in economics that have had an
impact on 'real life'. Let me just mention one example. In 1973, two American
economists, **Fischer Black** (1938–1995) and **Myron Scholes** (1941–), published a groundbreaking paper in which they provided an equation allowing one
to calculate the value of financial options. To do this, they made a number
of assumptions that looked natural. This paper started the modern field of
financial mathematics and led to the 1997 Nobel Memorial Prize in Economic
Sciences.[30]

The equation, now called the *Black–Scholes equation*, and its modifications
were applied to determine values of various financial derivatives (defined by the
International Monetary Fund (IMF) in a circular way as "financial instruments
that are linked to a specific financial instrument or indicator or commodity, and
through which specific financial risks can be traded in financial markets in their
own right"). This had a tremendous impact on the financial markets.

Unfortunately, Black and Scholes' approach was misused. One of their assumptions had been that the markets are random (that is, their movements are
not predictable). Warnings against this assumption by a number of researchers
(including Mandelbrot in his mentioned book, *The (Mis)behavior of Markets:
A Fractal View of Financial Turbulence*), did nothing to prevent the rise of
'risk management' companies, some of which took this assumption for granted
and assumed that risk can be quantified. In 2012, BBC News posted an article
with the ominous title *Black–Scholes: The maths formula linked to the financial crash*, discussing financial crises that could be linked to an inadvertent use
of the equation.[31]

Graph theory revisited One of the mathematical theories advanced in
the 20th century that lends itself to an informal
presentation is graph theory. I discussed it already in Chapter 6 in connection
with Euler's problem of the seven bridges in Königsberg. Graphs provide a
convenient way of representing information, which explains their wide use.

As a simple example, consider the problem of listing the compatibility of
the red blood cells. There are four blood types, A, B, AB, and O. Each blood
type can be given to recipients with the same blood type. The graph on the

next page explains the matter for different blood types. An arrow indicates the blood type can be given to recipients with another blood type. For example, type A blood can only be given to a person with blood type AB.

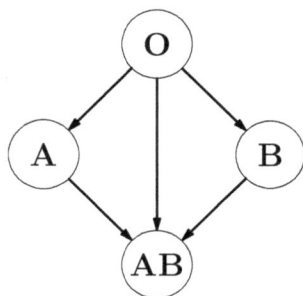

Blood compatibility chart.[32]

Graph theory has many useful applications. Once one succeeds to formalize a given problem by means of a graph, one can often use graph-theoretic techniques to solve it. An example is the *timetabling problem*, which in its simplest version, asks one to create a timetable for a given semester at a university. It involves assigning teachers to classes, each taught for a number of weeks, respecting natural constraints like each teacher being able to teach at most one course at the same time. Another example is the *Chinese postman problem*, which seeks to find a shortest route for a postman who has to deliver a post in a given city neighborhood. Yet another example of a graph-theoretic problem is the standard task of a GPS navigation system to find the shortest route between two locations.

Among many contributors to this area, two early ones need to be mentioned. **Frank Plumpton Ramsey** (1903–1930) was a brilliant English mathematician and economist who died prematurely at the age of 26. He was the founder of the *Ramsey theory*, which aims to prove that some 'order' within 'disorder' has to exist in graphs. The simplest example of a Ramsey-like result is the following statement: In any group of six people meeting at a party, either three of them know each other, or three of them are mutual strangers. Obviously, a more general version of this claim exists.[33]

Kazimierz Kuratowski (1896–1980) was a prominent Polish mathematician who contributed to many areas. In 1930, he established a famous theorem that determines when a graph is planar (recall that it means that it can be drawn in such a way that no two edges of it cross). This theorem became important in the production of integrated circuits since, to minimize construction costs, it is preferable to construct a given circuit as a planar graph or otherwise to find a layout with a minimal number of crosses, as each cross requires adding a bridge.

An example of a graph that is *not* planar is the one representing the gas, water, and electricity puzzle discussed in Chapter 6. In this graph, each of the three houses is connected to gas, water, and electricity. At the end of the 20th century, Kuratowski's theorem was vastly generalized by two American mathematicians, **Neil Robertson** (1938–) and **Paul Seymour** (1950–), in a series of papers totaling more than 500 pages.

Operations Research   Game theory and graph theory are just two subdisciplines in the field of Operations Research (OR).[34] It is a vast collection of scientific methods employed in the management of organizations that aims to find efficient and preferably optimal solutions to an array of practical problems. The vast increase in the scope of OR took place at the beginning of the Second World War when military supplies and personnel had to be moved efficiently.

Some problems earlier studied can be nowadays classified as OR problems. An example is the brachistochrone problem discussed in Chapter 6 (recall that it concerns finding the curve of the fastest descent between two points, one lying diagonally above the other), as it calls for finding an optimal solution.

Also, some problems involving probability are natural OR problems. For example, Poisson distribution, mentioned in the previous chapter, is nowadays widely used to make a forecast in the context of recurring processes, such as the arrival of customers at a store, the placing of orders, or incoming phone calls.

One of the most successful subdisciplines of OR is *linear programming*, which deals with optimal solutions to a set of linear inequalities. Some of the studied problems involve thousands of variables. The most broadly used algorithm is the *simplex method* developed by an American mathematician **George Dantzig** (1914–2005). Its optimized implementations allowed one to find optimal solutions in diverse areas such as healthy diet, transportation, physical distribution of warehouses, train timetable, or crew assignment, to name a few out of hundreds of applications.

Efficient algorithms   The advent of computers led to a growing interest in efficient algorithms, so that they can be implemented as usable computer programs. To design algorithms and analyze their efficiency, a whole array of techniques was developed during the last half a century. They draw on a number of mathematical areas, mainly combinatorics, algebra, probability, and logic.

One of the earliest examples of an algorithm is the *Euclidean algorithm* described in Euclid's *Elements*. It computes the greatest common divisor (*gcd*) of two natural numbers, so, the largest number that divides them without a remainder. For example, $gcd(39, 15) = 3$ because $39 = 3 \cdot 13$ and $15 = 3 \cdot 5$, and 13 and 5 have no common divisors bigger than 1. The algorithm is remarkably simple: just keep subtracting the smaller number from the larger one until both numbers become equal. For example, starting with the above pair (39,15), we get the following sequence of pairs:

$$(39, 15) \to (24, 15) \to (9, 15) \to (9, 6) \to (3, 6) \to (3, 3).$$

A more efficient version of this algorithm is incorporated into the RSA cryptosystem, while a more general version was incorporated into software systems that allow one to manipulate mathematical expressions.

Two more recent examples show how ubiquitous some algorithms are. In 1959, a Dutch computer scientist **Edsger Wybe Dijkstra** (1930–2002), published an algorithm now called the *shortest path algorithm* or *Dijkstra's algorithm*. Its purpose is to compute the shortest path in a graph in which each edge has some value associated with it.

Sorting is one of the most basic procedures needed to order elements according to some criterion (say, price, size, weight, or popularity). In 1961, British computer scientist **Tony Hoare** (1934–) proposed an elegant sorting algorithm that he called Quicksort.

Both algorithms are still widely used. Dijkstra's algorithm is indispensable in already mentioned GPS navigation systems for computing the shortest route between two locations, while Quicksort remains one of the most popular sorting algorithms. Both Dijkstra and Hoare received the Turing Award.

The P versus NP problem   Around the time when Dijkstra and Hoare proposed their algorithms, research was initiated—the aim of which was to quantify in more exact terms the concept of complexity of an algorithm. As a result, the notion of efficiency was made precise by appealing to the concept of computation by a Turing machine. This led to the rise of *computational complexity*, one of the most important areas in theoretical computer science.

In the early seventies, two researchers, American-Canadian **Stephen Cook** (1939–) and Russian-American **Leonid Levin** (1948–), independently established a crucial result in this area. It led to each of them formulating the most important open problem in theoretical computer science called the P *versus* NP problem. In the eighties, a letter from Gödel to von Neumann was discovered where he also stated this problem, albeit informally.[35] In turn, in 2012, the US National Security Agency (NSA) declassified a hand-written letter that John Nash sent to it in 1955. In this letter, he also stated this problem in the context of designing a safe cryptosystem.[36]

The problem is concerned with the question of whether computational tasks exist that are inherently difficult to solve, even though solutions to them are easy to check.[37] A couple of potential examples of such tasks can clarify the matter. Consider the following list of numbers:

$$14,175, \quad 15,055, \quad 16,616, \quad 17,495, \quad 18,072, \quad 19,390, \quad 19,731, \quad 22,161,$$
$$23,320, \quad 23,717, \quad 26,343, \quad 28,725, \quad 29,127, \quad 32,257, \quad 40,020, \quad 41,867,$$
$$43,155, \quad 46,298, \quad 56,734, \quad 57,176, \quad 58,306, \quad 61,848, \quad 65,825, \quad 66,042,$$
$$68,634, \quad 69,189, \quad 72,936, \quad 74,287, \quad 74,537, \quad 81,942, \quad 82,027, \quad 82,623,$$
$$82,802, \quad 82,988, \quad 90,467, \quad 97,042, \quad 97,507, \quad 99,564.$$

One can check that these numbers sum up to 2,000,000. The question is: can one break this list into two equal groups, such that each group sums up to 1,000,000? There is no known method to solve such a problem efficiently. On the other hand, for each splitting of the list into two groups, it is easy to check whether this is a solution.[38]

Next, given a list of cities and the distances between them, the problem is to find a shortest route that visits each city exactly once and returns to the point of origin. This problem, called the *Traveling Salesman Problem*, was first formulated in the 19th century. It is of obvious importance, for instance, in logistics: an early work on this problem was concerned with designing an efficient route of a school bus in a given district. A modified version of this problem asks for determination whether a route shorter than a given length exists such that it visits each city exactly once. Given a route, one can easily determine whether it satisfies both conditions. However, it is not known how to efficiently find such a route in general.

There is a general agreement that the question at the root of the P versus NP problem has a positive answer.[39] Otherwise, thousands of, at this stage, computationally difficult problems, in particular the above partition problem and the modified version of the Traveling Salesman Problem, would admit efficient solutions, which is highly unlikely. Informally, this belief is summarized as a statement that inventing solutions to challenging problems (for instance, producing a music masterpiece) is inherently more difficult than recognizing a solution (for instance, appreciating that Verdi's *Aida* is a masterpiece).

Applications in linguistics and computer science I already mentioned applications of ma-
thematics in economics (game theory), biology and meteorology (dynamical systems), and computer science (Turing machine). Applications of mathematics in linguistics were triggered by the works of **Noam Chomsky** (1928–), an American linguist, philosopher, and outspoken political analyst. In the second half of the 20th century, he spelled out his views on linguistics in a number of highly influential books and articles. What is relevant for this overview is that in his works, he introduced various categories of *formal grammar* that allows one to generate and analyze sentences formed in natural (or artificial) languages.

Subsequent advances in linguistics owe much to Chomsky's initial insights and to the development of *computer languages*, which are formal languages in which computer programs can be written. Initially, these languages were directly reflecting computer hardware by sticking close to its description in the form of simple instructions operating on registers and memory locations. In the late fifties, the first high-level programming language, FORTRAN, was de-signed and implemented. In this context, 'high-level' means that it enabled the programmers to express their programs in terms of more natural ('high level') constructs. The implementations of programming languages, called *compilers*, are crucially based on an efficient analysis of the employed formal grammar.

FORTRAN was the first of hundreds of programming languages that were created since then. Some of them attempt to stick as closely as possible to the basic mathematical concepts of sets, functions, and relations, though the programming languages that ignore such adherence rule supreme for reasons beyond the scope of this account.

One of the indirect consequences of Chomsky's work on formal grammars is the demise of the typesetter profession. In 1978, **Donald Knuth** (1938–), an American mathematician and computer scientist, a recipient of the Turing Award and the author of a renowned, still growing, multivolume monograph on algorithms and their mathematical analysis, released a free typesetting system TeX. Thanks to its versatility, TeX gradually superseded earlier typesetting systems used for mathematics. TeX allows one to produce high-quality books and articles with complex mathematical formulas, excellent drawings, figures, musical scores, representation of crossword puzzles, chess problems, and Babylonian numerals and Frege's forbidding notation, which are both used in this book. In particular, Wikipedia uses TeX to display mathematical formulas.

Nowadays, practically all physicists, mathematicians, and computer scientists write their articles in TeX or one of the systems built on top of it. All typesetting systems view the input text as a sentence in a specific formal language and transform it into a printable version.

Fourier analysis and wavelets   With the advent of computers, Fourier analysis became useful to describe and analyze periodic phenomena that arise in many disciplines. The list of application areas is impressive and includes astronomy, physics, meteorology, engineering, geology, and medicine. The crucial progress was achieved thanks to the *fast Fourier transform* (FFT) algorithm, which is one of the most important algorithms in computer science and engineering nowadays. We rely on it daily, as it is used in transmission through telephone lines, computer representation of images, computed tomography (CT) scanners, magnetic resonance imaging (MRI) scans, digital music (MP3), and voice recognition.

The 1979 Nobel Prize in Physiology or Medicine was awarded for "the development of computer assisted tomography", the purpose of which is to construct three-dimensional images from X-ray photos taken from different angles around the body.[40] In turn, the 1991 Nobel Prize in Chemistry was awarded for "the development of the methodology of high resolution nuclear magnetic resonance (NMR) spectroscopy", used to determine the structures of various chemical compounds.[41] Both works were crucially based on Fourier analysis.

Fourier's approach leads to a decomposition of a function into components, each with a precise frequency. This becomes an obstacle when one wants to model some phenomena in which both regular and irregular frequencies occur. This led to a search for alternative decompositions of a function that are better suited for some applications. Research on this subject was initiated in mathematics but continued by engineers, physicists, and computer scientists, and the subject became known as *wavelets*.

A breakthrough took place in 1987 when Belgian physicist and mathematician **Ingrid Daubechies** (1954–) found what is now called the *Daubechies wavelet*, which turned this approach into a practical tool that could be easily programmed.

In 1998, these techniques were successfully used in the Walt Disney Pictures and Pixar Animation Studios' computer-animated film *A Bug's Life*. But the most spectacular application of wavelets took place in 1992, when the FBI chose a wavelet-based method to compress its huge database of fingerprints. Nowadays, the JPEG2000 image compression standard uses wavelets.[42]

Three easy to explain results I already mentioned that most of the results obtained by mathematicians in the 20th century cannot be explained in such an informal account. Still, some results form a pleasing exception in the sense that it is at least easy to understand their consequences. Here are three examples.

- Imagine that you are in a hotel in Paris, and you lay on the floor a map of the city. There will then be exactly one point that corresponds to reality. This is a direct consequence of *Banach's fixed-point theorem*, established by a prominent Polish mathematician, **Stefan Banach** (1892–1945).
- Poincaré proved the *Hairy ball theorem*, informally explained as the statement: "You cannot comb a hairy ball flat without creating a cowlick".
  Both results are special cases of an important theorem in topology called *Brouwer's fixed-point theorem*, established by a Dutch mathematician, **L.E.J. (Bertus) Brouwer** (1881–1966).
- Another result established by Banach is the *Ham sandwich theorem*. It implies that one can always slice a ham sandwich with one cut so that the ham and both slices of bread are each divided into equal halves, no matter the size of all three ingredients.

Three famous results Additionally, three truly spectacular works that resulted in the solutions of famous problems can also be easily explained.

The first one is concerned with *map coloring*. A conjecture posed in the 19th century stated that every map could be colored using four colors in such a way that no two neighboring countries have the same color. Over the years, several faulty proofs were published. It was eventually proved in 1976 by **Kenneth Appel** (1932–2013) and **Wolfgang Haken** (1928–). An uncommon feature of the proof, which exceeded 100 pages, was that part of it was carried out using computer calculations. For this reason, some mathematicians initially refused to accept it. Later, a simplified version of the proof was produced, and the whole proof was checked using a computer program (a so-called *automated proof checker*). This is one of the most involved results in graph theory.

The second one, called *Fermat's last theorem*, goes back to Pierre de Fermat, who famously wrote in a Latin edition of Diophantus' *Arithmetica*:

"I have discovered a truly remarkable proof of this theorem which this margin is too small to contain." No other information has been found.

To state it, first, consider the equation $x^2 + y^2 = z^2$. It has solutions in the natural numbers. An example is $x = 3$, $y = 4$, $z = 5$. (Recall from Chapter 1 that such triples are called Pythagorean triples.) Fermat claimed that no solutions exist if we replace the power of 2 with a larger one; in other words, that for $n > 2$, the equation

$$x^n + y^n = z^n$$

has no solutions in natural numbers. This statement is called Fermat's Last Theorem.

The problem resisted attacks by several prominent mathematicians, including Euler and Gauss, who, over three centuries, succeeded only to prove some special cases. Then, in 1993, a British mathematician **Andrew Wiles** (1953–), after seven years of solitary work, provided a proof. The news reached the front page of the *New York Times*. Then a drama unfolded, as a gap was found in the proof. After an agonizing year, Wiles, with his former student, eventually succeeded to repair the proof and published the final version in 1995. The proof is more than 120 pages long and relies on techniques from algebra, geometry, and analysis.[43]

The third one is the *Kepler Conjecture*. It states informally that the most efficient way to pack oranges is to stack them in the obvious way used in fruit shops. In 1990, an involved purported proof of the conjecture appeared, but eventually, a consensus emerged that the proof was incomplete. The conjecture was finally established in 1998 by **Thomas Callister Hales** (1958–), who worked on it for 13 years. Part of the argument, running over 100 pages, makes use of—like in the case of the four colors theorem—a computer program. This led some mathematicians to doubt the validity of the proof. In response, Hales, with a number of computer scientists, launched a project aiming to check the whole proof by means of an automated proof checker. The project was successfully completed in 2015.

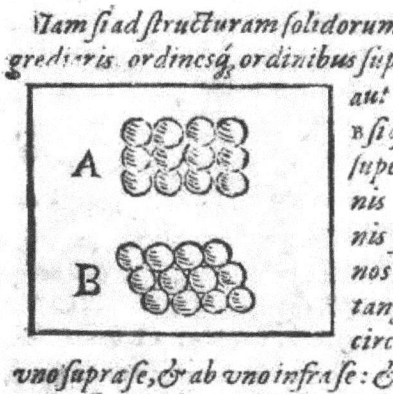

A drawing of Kepler conjecture from his 1611 book.[44]

Prizes Let me now discuss the most known prizes in mathematics. The most prestigious award is the *Fields medal*, bestowed every four years to two, three, or four mathematicians under the age of 40. It was established in 1936. In particular, over the years, five members of the Bourbaki group received the Fields medal.

Equally prestigious is the *Abel Prize*, established in 2002 by the Norwegian Academy of Science and Letters and modeled after the Nobel Prize. It does not have an age restriction and is awarded each year to one or two laureates 'to recognize pioneering scientific achievements in mathematics'. Among its winners are John Nash and Andrew Wiles.

Also, other generous prizes were established to help publicize mathematics in the eyes of the general public. In 1998, The Clay Mathematics Institute was founded by successful businessman and philanthropist Landon Clay and his wife. It announced, in 2000, seven *Millennium Prize Problems*, and to popularize them, offered one-million-dollar awards for solving each of them. One of these problems is the already mentioned Riemann hypothesis. Another one is the P versus NP problem. Then, in 2014, the annual *Breakthrough Prize in Mathematics* was announced, each worth three million dollars.

Some remarkable figures In addition to Nash, Turing, and Gödel, the 20th and 21st centuries featured another array of noteworthy and colorful mathematicians. One of them, a poor Indian clerk, **Srinivasa Ramanujan** (1887–1920), an autodidact in mathematics, single-handedly made several deep contributions to the number theory. He summarized some of them in a strangely worded letter to a renowned student of Bertrand Russell, **Godfrey Harold Hardy** (1877–1947), who recognized the genius of Ramanujan and eventually invited him from India to Cambridge. Their subsequent fruitful collaboration was interrupted by the premature death of Ramanujan, due to tuberculosis, at the age of 32. Ramanujan kept some of his findings in notebooks, without any proofs. Some of them were found only after his death. In particular, he provided several truly amazing and puzzling infinite sums that can be used to efficiently calculate π, like this one:[45]

$$\frac{1}{\pi} = \frac{2\sqrt{2}}{9801} \sum_{k=0}^{\infty} \frac{(4k)!(1103 + 26390k)}{(k!)^4 396^{4k}}.$$

It was only proved to be correct some 70 years later.

Ramanujan's impressive contributions to mathematics were honored by his early election as a Fellow of the Royal Society in London at the age of 29. In India, he was recognized mainly after his death. In 1962, an India Post stamp was issued bearing his photograph. By the 100th anniversary of his birth, three films about him were made in India. More recently, in the 21st century, a successful novel about his life appeared, and a well-received film about him was produced.[46]

Hardy, himself a brilliant mathematician who contributed to analysis and number theory, was a shy, elegant, and somewhat eccentric person with a keen interest in cricket. He once stated that his greatest contribution to mathematics was the discovery of Ramanujan. To the general public, Hardy is best remembered as the author of a charming little book, *A Mathematician's Apology*, which offers one of the most successful explanations of the beauty of mathematics in popular terms, and in which he defends the importance of pure mathematics, so devoid of any applications. The following quote is often cited:

> The mathematician's patterns, like the painter's or the poet's must be beautiful; the ideas, like the colours or the words must fit together in a harmonious way. Beauty is the first test: there is no permanent place in this world for ugly mathematics.

Another unusual person was a prominent Hungarian mathematician, **Paul Erdős** (1913–1996). From a certain point he had no permanent address, and until his death, while attending a conference in Warsaw, Poland, he traveled from one University to another with most of his belongings packed in his half-empty suitcase.

Erdős was one of the most prolific authors of the 20th century. In his long scientific career that started when he was 20, he wrote almost 1,500 papers and collaborated with more than 500 co-authors. He was always on the lookout for simple proofs, especially in geometry, number theory, and graph theory, and if he found one, he claimed that it was from 'the Book' in which God wrote the proofs for all theorems. Throughout his career, he offered financial awards for solving mathematical problems that he came up with, with amounts ranging between $25 and $10,000.

Paradoxically, Erdős was unwilling to accept that in the Monty Hall problem discussed in Chapter 6 (and also in Appendices 28 and 29), the participant should switch, and he was convinced only after being shown the outcomes of extensive computer simulations of randomly generated scenarios.[47]

Alexander Grothendieck (1928–2014), a Frenchman of German origins, was a member of the Bourbaki group. He is considered by many of his peers as the greatest mathematician of the 20th century. Grothendieck was a political activist, strongly critical of both Russian and American military operations. To protest the Vietnam War, he gave lectures 100 kilometers from Hanoi, during its bombing by the American forces. When he was awarded the Fields medal for his work in several fields, including algebraic geometry, he refused to accept it during the International Congress of the International Mathematical Union held in Moscow, to protest the arrest of two Russian writers.

After publishing a couple of thousand pages on deep generalizations of various mathematical theories, he suddenly burned 25,000 pages of his unpublished manuscripts and withdrew from public life to live his last 13 years of his life as a hermit in a secret location in the Pyrenees.[48]

Another reclusive mathematician is **Grigori Perelman** (1966–), who, after eight years of work, solved the *Poincaré conjecture* in 2003, one of the seven Millennium Prize problems initially posed 100 years earlier. He published his solution only in three short papers, which he posted on the Internet. The matters took an ugly turn when, in 2005, two Chinese mathematicians submitted for publication a 326-page paper in which they claimed that *they* finally established the Poincaré conjecture by building upon Perelman's incomplete arguments. The mathematical community thought otherwise and awarded Perelman the Fields medal in 2006. Perelman refused to accept it without announcing his reasons (unlike Jean-Paul Sartre, who declined the 1964 Nobel Prize in Literature in a long open letter). Also, he declined to receive the one-million-dollar prize of the Clay Mathematics Institute.[49]

William Thurston, another Fields medal winner, stated in 2010, in his laudation of Perelman:[50]

> Perelman, with tremendous focus and virtuosity, constructed a beautiful proof where I and others failed. [...] Perelman's aversion to public spectacle and to riches is mystifying to many. I have not talked to him about it and I can certainly not speak for him, but I want to say I have complete empathy and admiration for his inner strength and clarity, to be able to know and hold true to himself. Our true needs are deeper—yet in our modern society most of us reflexively and relentlessly pursue wealth, consumer goods and admiration. We have learned from Perelman's mathematics. Perhaps we should also pause to reflect on ourselves and learn from Perelman's attitude toward life.

Influence of the Internet   It is fitting to conclude this account of mathematics in the 20th and 21st centuries by pondering on the dramatic impact of the Internet, both on the way mathematics can be used and how mathematicians can work. In 2009, **Stephen Wolfram** (1959–), a precocious scientist who received his PhD in particle physics at the age of 20, launched a powerful query answering *WolframAlpha* system, https://www.wolframalpha.com/. In particular, it allows one to solve various equations, visualize functions, and compute answers to various purely numerical problems and mathematical queries formulated in an intuitive mathematical language (designed by Knuth for his TEX system). For example, it allows one to generate exact solutions to the third and fourth-degree equations that Italian mathematicians struggled with in the 16th century, to solve calculus problems considered by Leibniz and Newton, and to confirm the finding of Euler that $1 + \frac{1}{4} + \frac{1}{9} + \frac{1}{16} + \ldots = \frac{\pi^2}{6}$.

In turn, MathOverflow, http://mathoverflow.net/, is an interactive 'question and answer site for professional mathematicians' started in 2009. In 2018, its number of registered users exceeded 87,000. By now, more than 145,000 questions have been answered.

Further, some prominent scientists now maintain a blog that informs the public about recent developments in their disciplines. Perhaps the most influential blog in mathematics is the one run by **Terence Tao** (1975–), an Australian-born child prodigy (he attained his Master's degree at the age of 16) and recipient of the Fields medal and the Breakthrough Prize in Mathematics. The power of the Internet was further dramatized in 2010 by a massive, collective refereeing and eventual refutation of a purported solution of the P versus NP problem from the Millennium Prize list.[51]

The 21st century also saw a rise in collaborations made possible through the advanced use of the Internet. Most remarkably, in 2009, **Timothy Gowers** (1963–), a recipient of the Fields medal and the author of a wonderful little book, *Mathematics: A Very Short Introduction*,[52] introduced on his blog the idea of a *Polymath project*, a form of Internet-based collaboration whose purpose is to solve major mathematical problems. It led to a solution of some important problems, some of which were initiated by a discussion on Math-Overflow.

This overview of mathematics in the 20th and 21st centuries is not only informal but also incomplete. By necessity, several areas of mathematics were omitted, and several concepts, in particular, those introduced by discussed mathematicians (such as Hilbert space, von Neumann algebra, or Banach space), were left out.

Timeline

Alfred North Whitehead (1861–1947)
David Hilbert (1862–1943)
Bertrand Russell (1872–1970)
Henri Léon Lebesgue (1875–1941)

Godfrey Harold Hardy (1877–1947)
L.E.J. (Bertus) Brouwer (1881–1966)
Srinivasa Ramanujan (1887–1920)
Ronald Fisher (1890–1962)
Stefan Banach (1892–1945)
Norbert Wiener (1894–1964)
Kazimierz Kuratowski (1896–1980)
Alfred Tarski (1901–1983)
Frank Ramsey (1903–1930)
John von Neumann (1903–1957)
Andrey Nikolaevich Kolmogorov (1903–1987)
Kurt Gödel (1906–1978)
Alan Turing (1912–1954)
Paul Erdős (1913–1996)
George Dantzig (1914–2005)
Claude Shannon (1916–2001)
Edward Norton Lorenz (1917–2008)
Abraham Robinson (1918–1974)
Gérard Debreu (1921–2004)
Kenneth Arrow (1921–2017)
Benoit Mandelbrot (1924–2010)
Alexandre Grothendieck (1928–2014)
John Forbes Nash, Jr. (1928–2015)
Noam Chomsky (1928–)
Wolfgang Haken (1928–)
Edsger Wybe Dijkstra (1930–2002)
Kenneth Appel (1932–2013)
Paul Cohen (1934–2007)
Tony Hoare (1934–)
Vladimir Igorevich Arnold (1937–2010)
Fischer Black (1938–1995)
Neil Robertson (1938–)
Donald Knuth (1938–)
Stephen Cook (1939–)
Myron Scholes (1941–)
Leonid Levin (1948–)
Paul Seymour (1950–)
Andrew Wiles (1953–)
Ingrid Daubechies (1954–)

Thomas Callister Hales (1958–)
Stephen Wolfram (1959–)
Timothy Gowers (1963–)
Grigori Perelman (1966–)
Terence Tao (1975–)

Notes

[1] A.M. Odlyzko, Tragic loss or good riddance? The impending demise of traditional scholarly journals, *International Journal of Human-Computer Studies*, Volume 42(1), pp. 71–122, 1995.

[2] M. Kline, op. cit., p. 1077.

[3] Hilbert's address and the subsequent history of the mathematical problems on his list are discussed in B. Yandell, *The Honors Class: Hilbert's Problems and Their Solvers*, A K Peters/CRC Press, 2001.

[4] M. Kline, op. cit., p. 1003.

[5] This paradox is amusingly discussed in a graphic novel: A. Doxiadis and Ch. Papadimitriou, *Logicomix: An Epic Search for Truth*, Bloomsbury, 2009. It was preceded by five years by a less elementary paradox found by an Italian mathematician Cesare Burali-Forti.

[6] B. Russell, *My Philosophical Development*, Simon & Schuster, p. 86, 1959.

[7] Courtesy: Public Domain, Wikipedia Commons.

[8] An engaging recent biography of Gödel is S. Budiansky, *Journey to the Edge of Reason: The Life of Kurt Gödel*, Norton & Company, 2021.

[9] For an account of Tarski's life, see A.B. Feferman and S. Feferman, *Alfred Tarski: Life and Logic*, Cambridge University Press, 2004.

[10] The first programmable computer was built in 1941 by Konrad Zuse who was unaware of Turing's work.

[11] A photo taken from `https://www.flickr.com/photos/bankofengland/51001108075/sizes/l/`, available under a Creative Commons license.

[12] For an account of Turing's life, see A. Hodges, *Alan Turing: The Enigma*, Simon & Schuster, 1983.

[13] E.T. Bell, *The Development of Mathematics*, Dover Publications, p. 587, 1992.

[14] H. Steinhaus and S. Trybula, On a paradox in applied probabilities, *Bulletin of the Polish Academy of Sciences*, 7, pp. 67–69 (1959).

[15] An example of such a machine can be seen in a 10-second-long YouTube video 'Claude Shannon Ultimate Machine' at `https://youtu.be/G5rJJgt_5mg`.

[16] Shannon's life and work is extensively discussed in W. Poundstone, *Fortune's Formula*, Hill and Wang, 2006.

[17] For a recent biography of Wiener, see J.M. Almira, *Norbert Wiener: A Mathematician Among Engineers*, World Scientific, 2022.

[18] Courtesy: Public Domain, Wikipedia Commons.

[19] For a story of the Bourbaki project, see A.D. Aczel, *The Artist and the Mathematician*, High Stakes Publishing, 2006.

[20] S.H. Lui, An interview with Vladimir Arnol'd, *Notices of the AMS*, 44(4), pp. 432–438, 1997.

[21] For an account of this project, see S. Ornes, Researchers race to rescue the enormous theorem before its giant proof vanishes, *Scientific American*, 1 July 2015, `https://www.scientificamerican.com/article/researchers-race-to-rescue-`

the-enormous-theorem-before-its-giant-proof-vanishes.

[22]A. Shenitzer and J. Stillwell (eds.), *Mathematical Evolutions*, The Mathematical Association of America, p. 51, 2002. Quoted after D. Bressoud, op. cit., p. xi.

[23]Quoted after R. Siegmund-Schultze, Henri Lebesgue, in: T. Gowers, I. Leader, J. Barrow-Green (eds.), *The Princeton Companion to Mathematics*, Princeton University Press, p. 796, 2008.

[24]N.J. Rose, *Mathematical Maxims and Minims*, Rome Press Inc., p. 57, 1988.

[25]An open letter to the authors of calculus textbooks, https://math.vanderbilt.edu/schectex/ccc/gauge/letter/.

[26]The existence of such a model follows from standard results in mathematical logic. In this model, a positive number r is called *infinitesimal*, if for all natural numbers n, we have $r < \frac{1}{n}$.

[27]In a remarkable achievement, in 2007, Canadian computer scientist J. Schaeffer, after running a computer program almost nonstop for 18 years, determined that there is a strategy in checkers for each player that guarantees him a draw, see C. Thompson, Death of checkers, *New York Times*, 9 December 2007, http://www.nytimes.com/2007/12/09/magazine/09_15_checkers.html. In chess and Go, the search space is too large (at least for now) to draw such conclusions, but already in 1997, the computer program Deep Blue defeated the chess world champion G. Kasparov, while in 2016, the Google AlphaGo program won in Go against one of the strongest world players.

[28]See A. Bhattacharya, *The Man from the Future: The Visionary Life of John von Neumann*, Allen Lane, 2021, a recent biography of von Neumann and an account of his scientific contributions.

[29]This is an area with Fourier's work on heat as a primary example. It deals with specific equations in which unknowns are functions with more than one variable.

[30]The prize was awarded to R. Merton and M. Scholes "for a new method to determine the value of derivatives". Black died in 1995, so he could not be a recipient. Merton also contributed to the subject in a crucial way.

[31]Black–Scholes: The maths formula linked to the financial crash, BBC News, 28 April 2012, https://www.bbc.com/news/magazine-17866646. See also Chapter 17 of I. Stewart, *In Pursuit of the Unknown: 17 Equations That Changed the World*, Basic Books, p. 64, 2012, on which this article is partly based.

[32]Courtesy: Public Domain, Wikipedia Commons.

[33]To formulate the general version of the theorem proved by Ramsey, consider first the case of two groups of 7 people and 5 people, respectively. The theorem states that there exists a number n, such that in every group of n people, either 7 people know each other or 5 people are mutual strangers. Ramsey proved that this holds for any pair of numbers, not only 7 and 5.

There is also a version of the Ramsey theorem for infinite graphs.

[34]In Europe, usually the name of Operational Research is used.

[35]The letter is reproduced at https://rjlipton.wordpress.com/the-gdel-letter/.

[36]See S. Aaronson, P $\overset{?}{=}$ NP, https://www.scottaaronson.com/papers/pnp.pdf pp. 1, 3.

[37]More precisely, P stands for 'polynomial time' and NP for 'nondeterministic polynomial time'. The question is whether some problems cannot be solved in polynomial time (are inherently difficult to solve) but in nondeterministic polynomial time (solutions to them are easy to check).

[38]In fact, it is easy to check that the following split of this list yields a positive answer:

$15, 055$, $16, 616$, $19, 390$, $22, 161$, $26, 343$, $40, 020$, $41, 867$, $43, 155$, $46, 298$, $57, 176$, $58, 306$, $65, 825$, $66, 042$, $69, 189$, $74, 537$, $81, 942$, $82, 623$, $82, 988$, $90, 467$

and

$14, 175$, $17, 495$, $18, 072$, $19, 731$, $23, 320$, $23, 717$, $28, 725$, $29, 127$, $32, 257$, $56, 734$, $61, 848$, $68, 634$, $72, 936$, $74, 287$, $82, 027$, $82, 802$, $97, 042$, $97, 507$, $99, 564$.

Indeed, numbers in each group add up to 1,000,000. This example is taken from L. Fortnow, *The Golden Ticket: P, NP, and the Search for the Impossible*, Princeton University Press, 2013.

[39]See S. Aaronson, op. cit., for a discussion about the relevance of the problem and for a survey of its status in 2017.

[40]See https://www.nobelprize.org/prizes/medicine/1979/summary/.

[41]See https://www.nobelprize.org/prizes/chemistry/1991/summary/.

[42]For an accessible introduction to wavelets, see D. Mackenzie, Wavelets: Seeing the forest and the trees, National Academy of Sciences, http://www.nasonline.org/publications/beyond-discovery/wavelets.pdf, 2001.

[43]For a history of Fermat's Last Theorem, see S. Singh, *Fermat's Enigma: The Epic Quest to Solve the World's Greatest Mathematical Problem*, Anchor, 1998.

[44]Courtesy: Public Domain, Wikipedia Commons.

[45]Recall that for a number n, $n!$ denotes the product $1 \cdot 2 \cdot \ldots \cdot n$ and for an expression $e(k)$, for example

$$\frac{(4k)!(1103 + 26390k)}{(k!)^4 396^{4k}},$$

the expression $\sum_{k=0}^{\infty} e(k)$ stands for the infinite sum $e(0) + e(1) + e(2) + \ldots$. For example $\sum_{k=0}^{\infty} \frac{1}{k+1}$ stands for the infinite sum $1 + \frac{1}{2} + \frac{1}{3} + \ldots$.

[46]For a detailed account of Ramanujan's life that also extensively discusses Hardy's life, see R. Kanigel, *The Man Who Knew Infinity: A Life of the Genius Ramanujan*, Washington Square Press, 1992.

[47]For an account of Erdős' life, see, e.g., P. Hoffman, *The Man Who Loved Only Numbers: The Story of Paul Erdős and the Search for Mathematical Truth*, Hachette Books, 1998.

[48]For an account of Grothendieck's life, see A.D. Aczel, op. cit. and A. Jackson, Comme appelé du néant—as if summoned from the void: the life of Alexandre Grothendieck, parts I and II, Notices of the American Mathematical Society, 51(4), pp. 1038–1056 and 51(10), pp. 1196–1212, 2004.

[49]S. Nasar and D. Gruber, Manifold destiny. A legendary problem and the battle over who solved it, *New Yorker*, 28 August 2006, https://www.newyorker.com/magazine/2006/08/28/manifold-destiny. For an account of the history of the Poincaré conjecture, see G.G. Szpiro, *Poincaré's Conjecture*, A Plume Book, 2008.

[50]See http://www.claymath.org/perelman-laudations. I owe this quotation to P. Strzelecki, *Matematyka Współczesna dla Myślących Laików* (in Polish), Wydawnictwa Uniwersytetu Warszawskiego, 2011.

[51]For an account of this story, see A. Nazaryan, A most profound math problem, *New Yorker*, 2 May 2013, http://www.newyorker.com/tech/elements/a-most-profound-math-problem.

[52]T. Gowers, *Mathematics: A Very Short Introduction*, Oxford University Press, 2012.

Chapter 9

Final Remarks

One of the sobering conclusions of this short introduction is the visible absence of women in the history of mathematics. An account of the lives of a few known female mathematicians reveals their inferior social position throughout history. So far, only three women have been mentioned—Hypatia, Florence Nightingale, and Ingrid Daubechies.

The first woman who became known in mathematics after Hypatia was the Italian **Maria Gaetana Agnesi** (1718–1799), a child prodigy who mastered several languages, including Latin, Greek, and Hebrew, at the age of 13.[1] In 1748, she published a highly influential two-volume book that provided a systematic, modern treatment of algebra and analysis. In recognition of this work, she was appointed professor of mathematics at the University of Bologna two years later. At that time, however, she increasingly turned to religion and apparently never took up the position.

The first woman to receive a PhD in mathematics was the Russian **Sofia Kovalevsky** (1850–1891). At her time, women were not allowed to study in Universities in Russia and Germany. However, the ban in Germany did not apply to foreigners. This prompted Kovalevsky to arrange for a marriage of convenience that allowed her to leave Russia for Germany, where she eventually received a PhD at the University of Göttingen. After a turbulent life, she eventually became a professor of mathematics at the age of 39 at the University of Stockholm but died two years later.

Kovalevsky also published autobiographical novels depicting her life in Russia. This unusual combination of being both a mathematician and novelist inspired Alice Munro, the winner of the 2013 Nobel Prize in Literature, to write a moving novella about her.[2]

A more modern example of the inferior treatment of women in science is the case of **Emmy Noether** (1882–1935), a German mathematician who made key contributions to modern algebra and mathematical physics and who is regarded by many as the greatest woman mathematician. Hilbert's efforts to offer her employment at the mathematics department of the University of Göttingen encountered strong opposition from the University staff on the grounds that she was a woman. In spite of Hilbert's eloquent protest that "After all, we are a university, not a bath house", it took four years for Noether to obtain the unpaid position of a *privatdozent*.

However, in 1933, together with a number of colleagues, she lost her position because of her Jewish origins, which prompted her to migrate to the United States. After she passed away at the height of her career, Albert Einstein sent a letter to the *New York Times* in which he wrote:

> In the judgment of the most competent living mathematicians Fräulein Noether was the most significant creative mathematical genius thus far produced since the higher education of women began.

To change the perception that mathematics is 'only for men', the Association for Women in Mathematics was formed in 1971; it aims to encourage women to study mathematics and pursue careers in mathematical sciences. It sponsors, among others, an honorary lecture series named after Sofia Kovalevsky and Emmy Noether.

The only women who have so far received the Fields medal are the Iranian **Maryam Mirzakhani** (1977–2017), who received it in 2014 for her work on dynamical systems but passed away three years later and the Ukrainian **Maryna Sergiivna Viazovska** (1984–), who received it in 2022 for her work on Kepler conjecture in higher dimensions.

Some remarks on the nature of mathematics

To conclude this short account of the history of mathematics, let me briefly discuss the predominant views on its nature and subject.

The use of mathematics was initially limited to counting, measuring distances, and calculating simple areas. Nowadays, it is difficult to find a domain of science in which mathematics is not used. Some disciplines, like physics or astronomy, have relied on mathematics since their inception, while others, like biology and psychology, started to use mathematical methods and techniques only relatively recently.

One can argue that the history of mathematics reflects our progress in understanding the world. This belief in the universal importance of mathematics is reflected in various remarks made by physicists and philosophers.

I already mentioned Galileo's striking remark: "The book of nature is written in the language of mathematics." Newton's opinion that

> The latest authors, like the most ancient, strove to subordinate the phenomena of nature to the laws of mathematics.

shows that, in his view, mathematics was a key tool to understand the world surrounding us. Maxwell went even further. In 1870, he delivered a lecture for a general audience titled 'On the relations of mathematics to physics'. In it, he discussed differences between both fields but also stressed that each field influences the other.[3]

The reliance of physics on mathematics was elegantly captured by Albert Einstein, who began his 1949 essay, *The theory of relativity*, by stating:

> Mathematics deals exclusively with the relations of concepts to each other without consideration of their relation to experience. Physics too deals with mathematical concepts; however, these concepts attain physical content only by the clear determination of their relation to the objects of experience. This in particular is the case for the concepts of motion, space, time.[4]

A modern debate on the subject was triggered by Eugene Wigner, a Hungarian-American physicist and a Nobel laureate in Physics, who gave a lecture in 1959 titled 'Unreasonable effectiveness of mathematics in the natural sciences', which appeared the year after as an article.[5] In the discussion that followed, opinions ranged from a viewpoint that attributing such an universal importance to mathematics is an act of faith that amounts to ignoring many aspects of human experience (for instance, emotions), to a claim that the physical universe *is* mathematics.[6]

In the previous chapter concerned with mathematics in the 20th and 21st centuries, nothing was mentioned about the mathematical underpinnings of the special and general relativity theories of Einstein and of quantum mechanics, the main theories of 20th-century physics. They rely in a crucial way on various areas of mathematics, the discussion of which goes beyond the scope of a popular account. In fact, the role of mathematics in contemporary physics is so extensive that it deserves a separate book. How contemporary physics and mathematics are intertwined with each other is discussed in a popular recent book by Graham Farmelo.[7]

Philosophers and philosophically inclined mathematicians have been concerned with more fundamental matters regarding the nature and subject of mathematical truths. To understand the views prevalent in the 20th century, we need to go back to Plato. He argued that reality exists on two levels. One is the visible world, while the other one is embraced by his theory of Forms, according to which abstract objects exist. The latter viewpoint is usually referred to as *Platonism*.

Descartes stands at the origin of modern Platonism, according to which the truths of mathematics exist outside the material world and independently of our existence. This is also taken to be the case for mathematical objects, such as circles or sets. So, according to Descartes, mathematicians, when proving theorems, discover objective truths about these objects.

How we arrive at mathematical truths was notably the subject of Kant's philosophy. In his theory of knowledge, he made a distinction between *a priori* and *a posteriori* propositions. To justify the former, one does not rely on any experience, while a posteriori propositions do rely on experience. As an example of an a priori proposition, he mentioned the mathematical statement '$7 + 5 = 12$'.

To understand how we draw conclusions, he further introduced a distinction between *analytic* and *synthetic* propositions. The former ones refer to statements that are true by the nature of the definitions, for example, "All triangles have three sides". In contrast, the latter refer to our knowledge of the world. For example, the statement "All bodies are heavy" is synthetic because 'heavy' is not part of the definition of a body. Kant also crucially argued that the concepts of space and time are a priori, which makes it possible for us to draw conclusions about the world. Further, he viewed geometry and arithmetic theorems as a priori and synthetic. In particular, the statement '$7 + 5 = 12$' was, in his view, synthetic, since to determine its truth, we need to count using our fingers.

Modern Platonism was further developed by Frege. He agreed with Kant that geometry relies on our intuition of space but disagreed about the nature of arithmetic statements. In his book, *Die Grundlagen der Arithmetik*, Frege argued that the truths of arithmetic can be reduced to those of logic and, consequently, arithmetic theorems are a priori and analytic (where he referred to his own, broader, notion of an analytic statement). In particular, he argued that, in contrast to Kant, the statement '$7 + 5 = 12$' is analytic. To drive his point home, he took the statement '$135664 + 37863 = 173527$', the truth of which cannot be determined by counting with our fingers.

As already noted, the work of Frege influenced Russell, who endorsed his views on the nature of mathematical statements and argued that "mathematics [is] independent of us and our thoughts".[8] Another prominent proponent of modern Platonism was Gödel.

The opposite view, called *anti-Platonism*, was embraced among others by Henri Poincaré, who notably argued that in mathematics, a claim that some object 'exists' just means that its definition does not imply a contradiction. This view is closely related to *nominalism*, a philosophical viewpoint according to which abstract objects do not exist.

Logicism was previously mentioned in the context of Russell's and White-head's work on *Principia Mathematica*. It is the view that mathematics can be reduced to logic. Poincaré criticized logicism and found that mathematical ideas cannot be described in logical terms.

Yet another viewpoint in mathematics is called *constructivism*. It stipulates that mathematical proofs should be constructive in the sense that they should furnish an example whenever its existence is asserted. This view was beautifully captured by Hermann Weyl, who stated that a nonconstructive existence proof informs the world that a treasure exists without disclosing its location.[9] Weyl was a brilliant PhD student of Hilbert. In 1918, he published a monograph, *Das Kontinuum*, in which he initiated a program for the arithmetical foundations of mathematics. It was soon overshadowed by Hilbert's program; Hilbert was clearly aware of Weyl's book.

Constructivism is actually an umbrella name for a number of mathematical approaches that call for constructive proofs. One of its main proponents was the already mentioned Dutch mathematician Bertus Brouwer, who developed a form of constructivism that became known as *intuitionism*. In his view, mathematics is the activity of building mental constructions. So, the law of excluded middle (stating that A or not A holds for every proposition A) holds only if we either have a proof of A or provide an argument showing that any attempt at constructing a proof of A fails. This approach was further developed by a student of Brouwer and the already mentioned Russian mathematician Kolmogorov.

A modern twist on this discussion is furnished by previously mentioned research on formalizing mathematical proofs using automated proof checkers. These are computer programs that allow one to validate mathematical proofs by first presenting them as proofs in a logical system and, subsequently, by mechanically checking each step of this proof. The list of mathematical theorems whose proofs were verified in this way is spectacular and includes the fundamental theorem of algebra, mentioned in Chapter 7, the prime number theorem discussed in Chapter 7, Gödel's theorems, and, as already mentioned, the four colors theorem and the proof of Kepler conjecture. Some of the most successful automated proof checkers rely on intuitionism.

This debate about the nature of mathematics is by no means finished. It belongs more to the branch of philosophy called *philosophy of mathematics* than to mathematics itself.

Notes

[1]C.A. Pickover, *The Math Book*, Sterling, p. 180, 2009.

[2]A. Munro, *Too Much Happiness*, Douglas Gibson Books, 2009.

[3]G. Farmelo, *The Universe Speaks in Numbers*, Faber & Faber Ltd, pp. 36–40, 2019.

[4]A. Einstein, *The Theory of Relativity and Other Essays*, Philosophical Library/Open Road, 2015.

[5]E. Wigner, The unreasonable effectiveness of mathematics in the natural sciences, *Communications in Pure and Applied Mathematics*, vol. 13, pp.1–14, 1960.

[6]See M. Tegmark, *Our Mathematical Universe*, Alfred A. Knopf, 2014.

[7]G. Farmelo, op. cit.

[8]B. Russell, *Mysticism and Logic*, Doubleday Anchor Books, p. 65, 1957.

[9]The following result illustrates the point. We show that there exist irrational numbers x, y, such that x^y is rational. Recall that $\sqrt{2}$ is irrational. If $\sqrt{2}^{\sqrt{2}}$ is rational then take $x = y = \sqrt{2}$ and else take $x = \sqrt{2}^{\sqrt{2}}$ and $y = \sqrt{2}$, since $(\sqrt{2}^{\sqrt{2}})^{\sqrt{2}} = \sqrt{2}^2 = 2$. However, we do not know which of these two cases holds; that is, this proof does not construct the numbers x and y we are looking for. So, from the constructive point of view, this is not a satisfactory proof.

For the same reason, if one adopts the constructivist point of view, proofs by contradiction are not acceptable either.

Chapter 10

Further Reading

There are several outstanding web resources concerning the history of mathematics. I found particularly useful:

- The MacTutor History of Mathematics archive, `http://www-groups.dcs.st-and.ac.uk/~history`, set up by the School of Mathematics and Statistics of the University of St. Andrews in Scotland,
- The Wolfram MathWorld, `http://mathworld.wolfram.com` (advertised as "the web's most extensive mathematics resource"),
- The History of Mathematics website, `http://www.maths.tcd.ie/pub/HistMath/` by David R. Wilkins.

Also, I would like to recommend some attractive websites concerned with the history of mathematics:

- The Mathematical Association of America (MAA) has an online journal *Convergence* devoted to the history of mathematics. Its "Mathematical Treasures", `https://www.maa.org/press/periodicals/convergence/index-to-mathematical-treasures`, are "images of mathematical objects and of selected pages of mathematical manuscripts and texts from various libraries, museums, and private collections".
- The History of Mathematics project, `https://history-of-mathematics.org`, is a highly informative "virtual interactive exhibit being developed for the National Museum of Mathematics in New York City". It contains photos and descriptions of several fascinating ancient artifacts important for the history of mathematics, and relevant timelines.
- Interactive Mathematics Miscellany and Puzzles, `http://www.cut-`

the-knot.org/, by the late Alexander Bogomolny, which discusses (with proofs) several elementary problems in, among others, algebra, arithmetic, combinatorics, games, geometry, probability, and trigonometry. It contains several useful links and discussions.

- Theorem of the Day, https://www.theoremoftheday.org by Robin Whitty, that lists by now 266 theorems, each explained on one page, with appropriate links and helpful comments. The site also contains a wealth of links, in particular to an extensive bibliography of popular books on mathematics.

There are, of course, several books on the history of mathematics. However, hardly any of them is accessible for a general audience and most of them are brief on the 20th and 21st centuries. The following books, each written in a different style and requiring some basic familiarity with mathematics, are especially recommended:

- I. Stewart, *Taming the Infinite: The Story of Mathematics*, Quercus Publishing, 2009,
- A. Rooney, *The Story of Mathematics*, Arcturus, 2015,
- W.P. Berlinghoff and F.Q. Gouvêa, *Math through the Ages: A Gentle History for Teachers and Others*, 2nd edition, The Mathematical Association of America, 2015.

Ian Stewart, the author of the first book, wrote several other popular books on mathematics and specific aspects of its history. Especially informative are:

- I. Stewart, *In Pursuit of the Unknown: 17 Equations That Changed the World*, Basic Books, 2012,
- I. Stewart, *Significant Figures: The Lives and Work of Great Mathematicians*, Basic Books, 2017.

John Tabak published a series of highly readable, popular books on history of mathematics, in particular

- J. Tabak, *Numbers: Computers, Philosophers, and the Search for Meaning*, Facts on File, 2004,
- J. Tabak, *Algebra: Sets, Symbols, and the Language of Thought*, Facts on File, 2004,
- J. Tabak, *Geometry: The Language of Space and Form*, Facts on File, 2004,

- J. Tabak, *Probability and Statistics: The Science of Uncertainty*, Facts on File, 2004.

For readers interested in pursuing these matters further, here are some pointers to extensive (the shortest one counts 688 pages) and highly readable histories of mathematics, in the increasing order of difficulty:

- U.C. Merzbach and C.B. Boyer, *A History of Mathematics*, 3rd edition, John Wiley & Sons, 2011,
- D.M. Burton, *The History of Mathematics: An Introduction*, 7th edition, McGraw-Hill Science, 2011,
- I. Grattan-Guinness, *The Rainbow of Mathematics*, W.W. Norton & Company, 2000,
- V.J. Katz, *A History of Mathematics, an Introduction*, Addison-Wesley, 3rd edition, 2009,
- M. Kline, *Mathematical Thought From Ancient to Modern Times*, Oxford University Press, 1972.

Those with less patience might study

- D.J. Struik, *A Concise History of Mathematics*, Dover Publications, 4th edition, 1987,

which is only 228 pages long, though it assumes a reasonable knowledge of mathematics. Mathematics in 20th century is discussed in more detail in

- P. Odifreddi, *The Mathematical Century: The 30 Greatest Problems of the Last 100 Years*, Princeton University Press, 2004,
- K. Devlin, *Mathematics: The New Golden Age*, Pelican Books, 1988,
- I. Stewart, *Great Problems of Mathematics*, Profile Books Ltd, 2014,

where the last two references provide an accessible account of some important mathematical developments in the last 50 years.

For those interested in reading about history of mathematics *together* with readable proofs, the following books are strongly recommended:

- T. Danzig, *Number: The Language of Science*, Plume, 2007 (originally published in 1930),
- R. Courant and H. Robbins, *What is Mathematics?*, Oxford University Press, 2009 (originally published in 1941; revised by I. Stewart),

- J. Stillwell, *Elements of Mathematics: From Euclid to Gödel*, Princeton University Press, 2016.

Further,

- T. Gowers, I. Leader, and J. Barrow-Green (eds.), *The Princeton Companion to Mathematics*, Princeton University Press, 2008

is a marvelous (though advanced) guide to mathematics that discusses, in an accessible way, the most relevant mathematical concepts and theorems, the main areas of mathematics, and offers short biographies of the most important mathematicians.

Finally, let me also mention three novels in which the main character is a real mathematician:

- L. Infeld, *Whom the Gods Love: The Story of Évariste Galois*, Whittlesey House, 1948: a novel about Évariste Galois,
- D. Leavitt, *The Indian Clerk*, Bloomsbury, 2007: a novel about Srinivasa Ramanujan,
- D. Kehlmann, *Measuring the World*, Vintage Books, 2007: a novel about Carl Friedrich Gauss,

three novels in which the main character is a fictitious mathematician:

- P. Schogt, *The Wild Numbers*, Plume, 2001,
- A. Doxiadis, *Uncle Petros and Goldbach's Conjecture*, Bloomsbury, 2001,
- Y. Ogawa, *The Housekeeper and the Professor*, Picador, 2009,

and three films about mathematicians that were all very well received:

- *A Beautiful Mind* from 2001, directed by Ron Howard, about the life of John Nash, inspired by the book: S. Nasar, *A Beautiful Mind*, Simon & Schuster, 1998,
- *The Imitation Game* from 2014, directed by Morten Tyldum, about the life of Alan Turing, inspired by the book: A. Hodges, *Alan Turing: The Enigma*, Simon & Schuster, 1983,
- *The Man Who Knew Infinity* from 2015, directed by Matthew Brown, about the life of Srinivasa Ramanujan, inspired by the book: R. Kanigel, *The Man Who Knew Infinity: A Life of the Genius*, C. Scribner's, 1991.

Appendices

1 A proof of Thales' theorem

Thales' theorem mentioned in Chapter 2 states the following.

Suppose that we inscribe a triangle into a circle in such a way that one of its sides is the diameter of the circle. Then the angle opposite the diameter side is right. In the picture below, we claim then that $\gamma = 90°$.

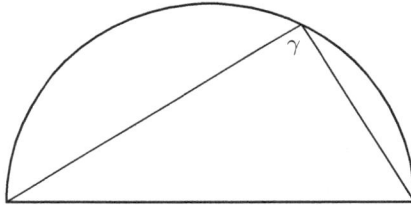

Thales' theorem.

To prove it, we connect the vertices of the triangle with the center of the circle and mark the relevant angles as in the picture below.

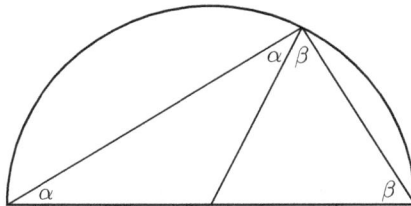

Proof of Thales' theorem.

We now use the observation that in each triangle, the angles opposite equal sides are equal. This justifies the double occurrences of α, since the sides opposite both αs are equal to the radius of the circle and analogously with the βs. The sum of the angles in each triangle is $180°$, so we get $\alpha+(\alpha+\beta)+\beta = 180°$. Hence, $\gamma = \alpha + \beta = 90°$.

2 Three proofs of the Pythagorean theorem

Perhaps a simplest proof

The following proof was rediscovered several times. It is sometimes attributed to Bhaskara II, an Indian mathematician discussed in Chapter 4.[1] Consider the following drawing.

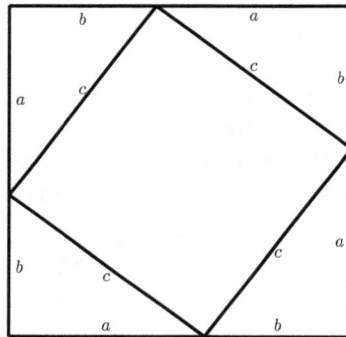

A proof of the Pythagorean theorem.

The area of the large square is $(a + b)^2$. It is covered by a square with the side c and four right-angled triangles, each with sides a, b, and c. So we get

$$(a + b)^2 = c^2 + 4\frac{ab}{2},$$

or

$$a^2 + 2ab + b^2 = c^2 + 2ab,$$

which yields the conclusion.

A proof suggested by a drawing in Zhou Bi Suan Jing

In Chapter 3, I mentioned a drawing that appeared in an ancient Chinese mathematical text, *Zhou Bi Suan Jing*. Consider now this drawing together with a figure in which we mark the relevant sides by a, b, and c.

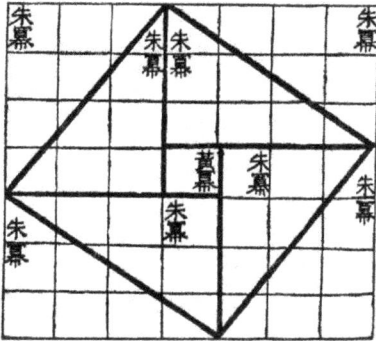

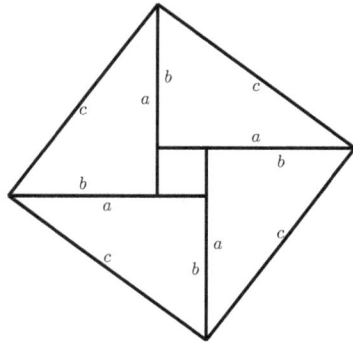

A drawing from *Zhou Bi Suan Jing* and a proof of the Pythagorean theorem.

The area of the large square in the figure on the right is c^2. It is covered by four right-angled triangles, each with the sides a, b, and c, and a square with the side $a - b$. So we get

$$4\frac{ab}{2} + (a - b)^2 = c^2,$$

or

$$2ab + a^2 - 2ab + b^2 = c^2,$$

which yields the conclusion.

Garfield's proof

Consider the following drawing of a trapezoid constructed using two copies of the original right-angled triangle with the side lengths a, b, and c, and a right-angled isosceles triangle with the congruent size of length c:

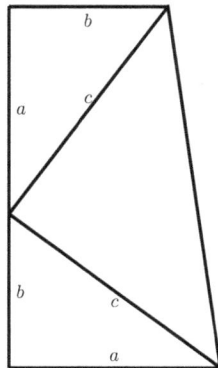

The area of the trapezoid equals its height times the average of the bases, so $(a+b)\frac{a+b}{2}$, i.e., $\frac{1}{2}(a+b)^2$, which is $\frac{a^2}{2} + ab + \frac{b^2}{2}$. But it also equals the sum of the areas of the three triangles, that is, $2\frac{ab}{2} + \frac{c^2}{2}$, i.e., $ab + \frac{c^2}{2}$. So $\frac{a^2}{2} + ab + \frac{b^2}{2} = ab + \frac{c^2}{2}$, from which $a^2 + b^2 = c^2$ follows.

3   The broken bamboo problem

Recall the broken bamboo problem mentioned in Chapter 3.

> A 10-chi-high bamboo broke, and its top touches the ground 3 chi from the base of the stem. At what height did it break?

Consider the original drawing side by side with the one of a relevant right-angled triangle.

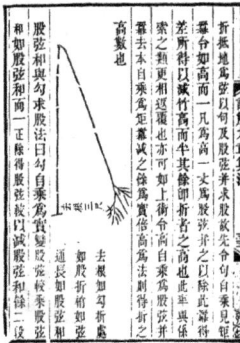

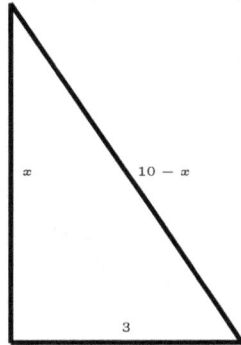

The broken bamboo problem and its representation.

Denote the length of the broken part by x. By the Pythagorean theorem, we have $x^2 + 3^2 = (10-x)^2$, or $x^2 + 9 = 100 - 20x + x^2$, which yields $x = \frac{91}{20} = 4.55$. So the bamboo was broken at the height of 4.55 chi.

4   Irrationality of $\sqrt{2}$ and the proportions of the A4 paper sheet

Consider a positive number d such that $d^2 = 2$. It is denoted by $\sqrt{2}$. We prove that $\sqrt{2}$ is irrational. Suppose by contradiction that $\sqrt{2}$ is rational, that is, $\sqrt{2} = \frac{p}{q}$ for some natural numbers p and q. Among all such pairs p, q, choose the one with the smallest p. By squaring both sides, we get $2 = \frac{p^2}{q^2}$, i.e., $2q^2 = p^2$. So p must be even, say $p = 2r$. Hence, $2q^2 = 4r^2$, i.e., $q^2 = 2r^2$.

So q must be even, as well, say $q = 2s$. Hence, $\frac{p}{q} = \frac{r}{s}$, which contradicts the choice of the pair p, q.

In connection with $\sqrt{2}$ I mentioned in Chapter 2 on page 12, the A4 paper sheet. Reconsider its drawing with the sides a and b, where $a > b$. By folding it in half along its longer edge, we get the A5 paper sheet with the sides b and $\frac{a}{2}$, where $b > \frac{a}{2}$. The proportions of the A4 and A5 sides are supposed to be the same, that is, we stipulate that

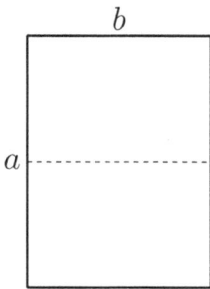

$$\frac{a}{b} = \frac{b}{\frac{a}{2}}.$$

The A4 paper sheet.

Thus, $\frac{a^2}{2} = b^2$, which is the same as $\frac{a^2}{b^2} = 2$, and this, in turn, is equivalent to $\frac{a}{b} = \sqrt{2}$.

5 Three proofs that there are infinitely many prime numbers

Call a divisor of a natural number *proper* if it is larger than 1. Below we shall repeatedly use the observation that for a natural number greater than 1, its smallest proper divisor is a prime number.

Original proof

The original proof by Euclid is as follows. It suffices to prove that for every finite set of prime numbers, there is a prime number not belonging to it.

So consider an arbitrary finite set of prime numbers. Take the product r of these prime numbers. Then, either $r + 1$ is a prime number, or it is divisible by a prime number p. In the latter case, p is not from the original set, as each number from this set divides r, so it does not divide $r + 1$. So, in both cases, we produced a prime number not in the original set.

The case analysis can be shortened by taking as a new prime number the smallest proper divisor of $r + 1$.

A variation on the original proof

It suffices to prove that for every natural number n, there is a prime number p larger than n. Recall that $n!$ denotes the product $1 \cdot 2 \cdot \ldots \cdot n$. We claim that we can choose for p the smallest proper divisor of $n! + 1$. It suffices to prove that no number smaller than $n + 1$ is a proper divisor of $n! + 1$. But every such number divides $n!$, so it does not divide $n! + 1$.

A recent proof

The following proof is due to F. Saidak.[2] For a natural number n, denote by $\omega(n)$ the set of its prime divisors. We repeatedly use the observation that two consecutive natural numbers have no common proper divisors. Hence, for every natural number $n \geq 1$

- the set $\omega(n(n+1))$ is a union of the sets $\omega(n)$ and $\omega(n+1)$,
- the smallest proper divisor of $n+1$ is not in the set $\omega(n)$.

So, the set $\omega(n(n+1))$ includes the set $\omega(n)$ and contains a prime number that is not in the set $\omega(n)$.

This means that starting with $n_1 = 1$, the infinite sequence of numbers

$$n_2 = n_1(n_1 + 1), \; n_3 = n_2(n_2 + 1), \; n_4 = n_3(n_3 + 1), \ldots,$$

produces an infinite sequence of strictly growing sets of prime numbers

$$\omega(n_2), \; \omega(n_3), \; \omega(n_4), \ldots.$$

Hence, the set of prime numbers is infinite.

6 Eratosthenes' estimate of the Earth's circumference

Eratosthenes reasoning was as follows. From the length of the caravan travels from Alexandria to Syene (now Aswan), he concluded that the distance between two cities equals 5,000 *stades*, an ancient distance measure. He assumed that both cities have the same longitude (which is not exactly true). He knew that at high noon of a specific day (the summer solstice), the Sun shines right above Syene.

On the same day, he measured at high noon the angle at which the Sun shines in Alexandria (which was 7.2 degrees off the zenith), for example, using some obelisk. Further, he assumed—because of the huge distance between the Sun and the Earth—that the sunrays in both cities at the same time were parallel. The latter assumption allowed him to conclude that 7.2 degrees was also the curvature of the Earth between the two cities; see the drawing below.[3]

Since $\frac{360}{7.2} = 50$, he multiplied 5,000 by 50 and obtained the result of 250,000 stades, which is approximately 40,000 kilometers. There are some (small) inaccuracies in his reasoning. Further, historians differ in their views of the exact length of a stade, which means that one considers it possible that Eratosthenes overestimated the circumference of the Earth by 10–15%.

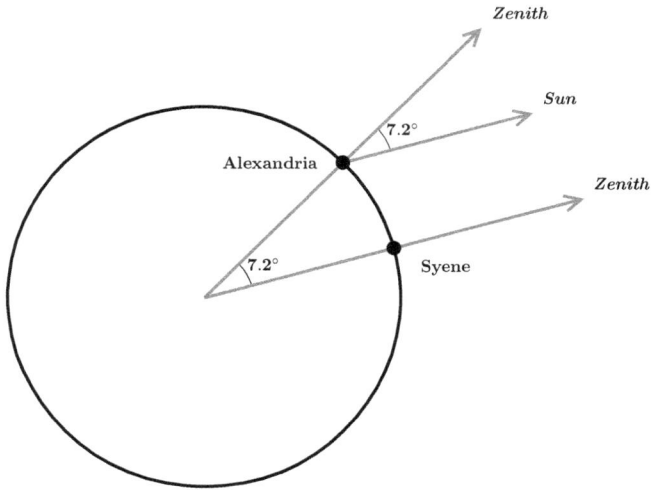

An illustration of Eratosthenes' argument.

7   A proof of Ptolemy's theorem

Ptolemy's theorem was introduced in Chapter 2 on page 22. The following proof of it, due to W. Derrick and J. Herstein, was presented as a proof without words, with just two drawings, the first one and the last one.[4] To explain the details, we added one more drawing.

Recall that a *peripheral angle* at a point on the circle is the angle subtended by a chord of the circle. Note that Thales' theorem implies that the peripheral angles subtended by a given diameter of the circle are equal (namely, they are all right angles). The following generalization of this result can be proved in a similar way.

Theorem The peripheral angles subtended by a given chord of the circle are equal.

Mark now the appropriate angles in the original drawing of the quadrilateral, as on the next page. The double occurrences of α, β, γ, and δ are justified by the abovementioned theorem. For example, both peripheral angles marked by α are subtended by side a.

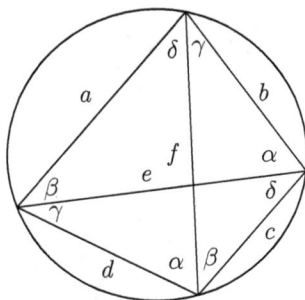

To prove the theorem (that is, that $ac + bd = ef$), we now select from the original drawing the following three triangles that we appropriately rotate.

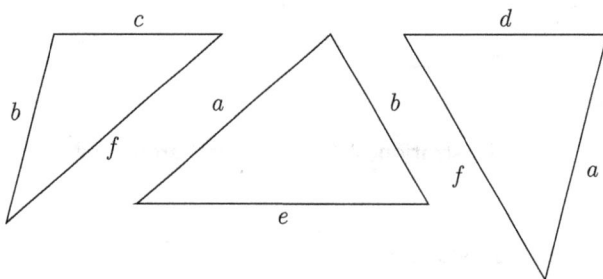

Next, we scale them respectively by a, f, and b so that they can be put side by side to form the figure below, in which we reuse the original markings of the angles.

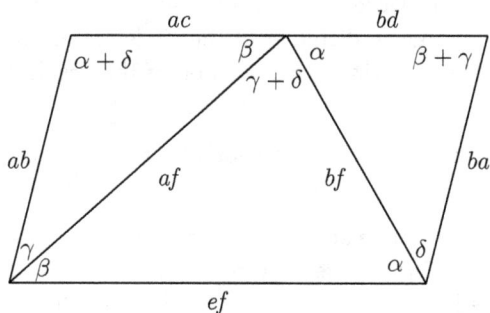

Note that ac and bd lie on the same line because $\beta + \gamma + \delta + \alpha = 180°$, as the sum of the angles of the original quadrilateral is $2(\alpha + \beta + \gamma + \delta) = 360°$. So the above figure is a quadrilateral.

The angle markings also reveal that the pairs of the opposite angles of the quadrilateral are equal: they are both $\alpha + \delta$ or both $\beta + \gamma$. So this quadrilateral is a parallelogram, and hence $ac + bd = ef$.

8 Al-Biruni's computation of the radius of the Earth

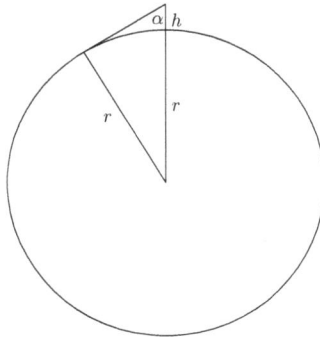

Al-Biruni's computation of the radius of the Earth.

Al-Biruni proceeded as follows. First, he computed the height h of a hill by comparing the angles subtended by it at two locations with a known distance apart. (This can be done by a simple trigonometric argument that we omit.) Then, using the astrolabe, he computed the angle α between the tangent and the vertical line from the top of the mountain to the horizon. Let r denote the radius of the circle. Given that

$$\sin(\alpha) = \frac{r}{r + h},$$

we then get

$$r = \frac{h \sin(\alpha)}{1 - \sin(\alpha)}.$$

9 Oresme's proof of divergence

In Chapter 4, I mentioned on page 46 that Nicolas Oresme established that the sum

$$1 + \tfrac{1}{2} + \tfrac{1}{3} + \tfrac{1}{4} + \cdots$$

diverges. His proof was geometric, but it can easily be explained by an algebraic argument. Note that

$$1 + \tfrac{1}{2} + \tfrac{1}{3} + \tfrac{1}{4} + \tfrac{1}{5} + \tfrac{1}{6} + \tfrac{1}{7} + \tfrac{1}{8} + \cdots$$
$$= 1 + \tfrac{1}{2} + (\tfrac{1}{3} + \tfrac{1}{4}) + (\tfrac{1}{5} + \tfrac{1}{6} + \tfrac{1}{7} + \tfrac{1}{8}) + \cdots$$
$$> 1 + \tfrac{1}{2} + (\tfrac{1}{4} + \tfrac{1}{4}) + (\tfrac{1}{8} + \tfrac{1}{8} + \tfrac{1}{8} + \tfrac{1}{8}) + \cdots$$
$$= 1 + \tfrac{1}{2} + \quad \tfrac{1}{2} \quad + \qquad \tfrac{1}{2} \qquad + \cdots$$

So successively grouping larger sections of the infinite sum, we exceed arbitrary long repetitions of $\tfrac{1}{2}$, which implies divergence of the original infinite sum.

10   A minimal introduction to complex numbers

Assume that the equation $x^2 = -1$ has a solution, $\sqrt{-1}$, that we subsequently denote by i. So, by definition, $i^2 = -1$. Then, we can represent the entity $\sqrt{-b}$, where b is a positive real number, by $i\sqrt{b}$, since $(i\sqrt{b})^2 = i^2 b = -b$.

We call the entities of the form $a + ib$ *complex numbers*. It is easy to extend the addition and subtraction operation to complex numbers by applying the laws of arithmetic:

$$(a + ib) + (c + id) = (a + c) + i(b + d),$$
$$(a + ib) - (c + id) = (a - c) + i(b - d).$$

Further, since $i^2 = -1$, we also can define the multiplication of two complex numbers, again applying the laws of arithmetic:

$$(a + ib) \cdot (c + id) = ac + iad + ibc + i^2 bd = (ac - bd) + i(ad + bc).$$

To define the division $\frac{a+ib}{c+id}$ (which is actually not used in the sequel), we multiply the numerator and denominator by $c - id$ and use the definition of multiplication (we assume here that $c + id \neq 0$, i.e., that $c \neq 0$ or $d \neq 0$, which is equivalent to the condition $c^2 + d^2 \neq 0$):

$$\frac{a + ib}{c + id} = \frac{(a + ib)(c - id)}{(c + id)(c - id)} = \frac{(ac + bd) + i(bc - ad)}{c^2 + d^2} = \frac{ac + bd}{c^2 + d^2} + i\frac{bc - ad}{c^2 + d^2}.$$

After this introduction to complex numbers, let us return to the formula considered in the 16th century by Bombelli (see page 54):

$$\sqrt[3]{2 + \sqrt{-121}} + \sqrt[3]{2 - \sqrt{-121}}.$$

We can rewrite it as

$$\sqrt[3]{2 + 11i} + \sqrt[3]{2 - 11i}.$$

Let us now check that $2 + i = \sqrt[3]{2 + 11i}$ by evaluating $(2 + i)^3$, which should yield $2 + 11i$. We use the formula

$$(a + b)^3 = a^3 + 3a^2b + 3ab^2 + b^3. \tag{1}$$

Noting that $i^2 = -1$ and $i^3 = -i$, we obtain

$$(2 + i)^3 = 2^3 + 3 \cdot 2^2 i + 3 \cdot 2 \cdot i^2 + i^3 = 8 + 12i - 6 - i = 2 + 11i.$$

By analogous calculations $(2 - i)^3 = 2 - 11i$. We conclude that

$$\sqrt[3]{2 + \sqrt{-121}} + \sqrt[3]{2 - \sqrt{-121}} = 2 + i + 2 - i = 4,$$

as claimed by Bombelli.

11 A derivation of the solutions of quadratic equations

We discuss the solutions of quadratic equations only here since they involve complex numbers. We follow an intuitively appealing approach here, recently found by an American mathematician Po-Shen Loh.[5]

Assume that a, b, c are some reals such that $a \neq 0$. (The same presentation works when a, b, c are complex numbers.) We want to solve the equation

$$ax^2 + bx + c = 0. \tag{2}$$

We first simplify it to

$$x^2 + px + q = 0$$

by introducing

$$p = \frac{b}{a} \text{ and } q = \frac{c}{a}.$$

Both equations have the same solutions.

If two numbers A and B satisfy the equality

$$x^2 + px + q = (x - A)(x - B),$$

then the left-hand side becomes 0 precisely when $x = A$ or $x = B$.

To find A and B, note that the above equality implies $A + B = -p$. So, for some z, we have

$$A = -\frac{p}{2} + z \text{ and } B = -\frac{p}{2} - z. \tag{3}$$

Hence

$$q = A \cdot B = \frac{p^2}{4} - z^2.$$

Consequently

$$z = \sqrt{\frac{p^2}{4} - q} \text{ or } z = -\sqrt{\frac{p^2}{4} - q},$$

from which by replacing p and q by its values and noting that

$$\frac{p^2}{4} - q = \frac{b^2}{4a^2} - \frac{c}{a} = \frac{b^2 - 4ac}{4a^2}$$

we conclude from equation (3) that

$$A = -\frac{b}{2a} + \frac{\sqrt{b^2 - 4ac}}{2a} \text{ and } B = -\frac{b}{2a} - \frac{\sqrt{b^2 - 4ac}}{2a}.$$

This is almost the same as what is being taught in secondary school, where one first introduces $\Delta = b^2 - 4ac$ and distinguishes three cases, $\Delta > 0$, $\Delta = 0$, $\Delta < 0$, which yield, respectively, 2, 1, or 0 solutions. In the above presentation, this case distinction disappears because we implicitly allow solutions in complex numbers.

Note that if $\Delta = b^2 - 4ac < 0$, then equation (2) has two solutions in complex numbers, namely

$$A = -\frac{b}{2a} + i\frac{\sqrt{-\Delta}}{2a} \text{ and } B = -\frac{b}{2a} - i\frac{\sqrt{-\Delta}}{2a}.$$

12   Golden ratio and the pentagram

The golden ratio was mentioned on page 13. I discuss it only here since its analysis involves solving a quadratic equation. It is defined as follows. Divide an interval a into two parts, b and c, such that $b \leq c$. Consider now two fractions, $\frac{a}{c}$ and $\frac{c}{b}$. If they are equal, then, by definition, they are equal to the golden ratio.

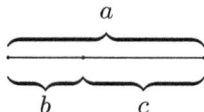

a

b c

Golden ratio: $\frac{a}{c} = \frac{c}{b}$.

To compute it, consider alternatively an interval divided into two parts, with respective lengths 1 and ϕ, where $1 \leq \phi$. If $\frac{1+\phi}{\phi} = \frac{\phi}{1}$, then ϕ is the golden ratio. This equality yields the equation

$$\phi^2 - \phi - 1 = 0.$$

We solve it using the formulas given in Appendix 11. Here

$$\Delta = (-1)^2 - 4 \cdot 1 \cdot (-1) = 5,$$

so we get two solutions

$$\phi = \frac{1 + \sqrt{5}}{2} \text{ and } \phi = \frac{1 - \sqrt{5}}{2}.$$

The second one yields a negative result, so needs to be discarded. Since $\sqrt{5}$ is irrational, ϕ is irrational, as well.

The letter ϕ is used as a reference to Phidias, one of the greatest sculptors of Ancient Greece, who apparently used the golden ratio in his architectural design.

Let us now return to the pentagram mentioned in Chapter 2 and discuss its connection with the golden ratio. Consider the following drawing of a pentagram to which the sides of the corresponding regular pentagon are added. The relevant sides of the pentagon are denoted by a, and the relevant diagonals (so the sides of the original pentagram) by b.

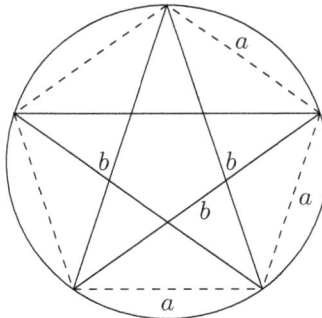

A pentagram and a regular pentagon.

By Ptolemy's theorem, discussed in Appendix 7, applied to the quadrilateral with the sides a, a, a, and b (and diagonals b and b), we have

$$a^2 + ab = b^2.$$

Dividing both sides by a^2, we get

$$1 + \frac{b}{a} = \left(\frac{b}{a}\right)^2.$$

This means that $\frac{b}{a}$ satisfies the same equation as the golden ratio, i.e., $\frac{b}{a} = \phi$.

13   A construction of a regular pentagon

The following elegant construction of a regular pentagon is due to Yosifusa Hirano, a 19th-century Japanese mathematician.[6] It is based on the following simple drawing.

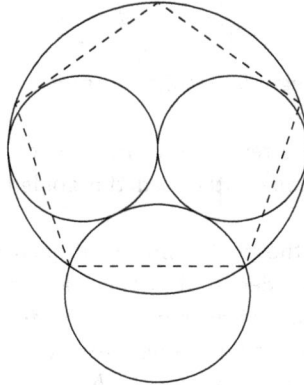

Hirano's construction of a regular pentagon.

So we first draw a circle. Next, we draw two circles inside, each half of its size, so that they touch each other and the original circle. Finally, we draw the fourth circle so that it touches the smaller circles. Its center is located at the intersection of the first circle with its diameter that is perpendicular to the diameter intersecting the centers of the smaller circles. The claim is that the intersection points of the first and last circles determine the side of the pentagon inscribed in the first circle.

To prove the correctness of this construction, we first consider an isosceles triangle in which the duplicated side b is in the golden ratio ϕ to its base a (such a triangle is called a *golden triangle*). We show that its angles are $36°, 72°$, and $72°$.

We first mark the distance a at one duplicated side, connect the obtained point with the opposite vertex, and name the new line and the relevant angles, as in the diagram below.

By assumption, $\frac{b}{a} = \phi$. Since ϕ satisfies the equation $\phi^2 - \phi = 1$, we have $\phi = \frac{1}{\phi-1}$, so

$$\frac{b}{a} = \frac{1}{\frac{b}{a} - 1} = \frac{a}{b - a} = \frac{a}{c}.$$

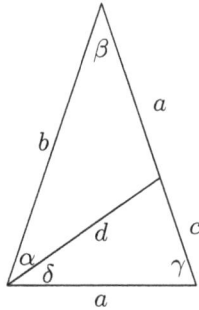

A golden triangle.

Consider now the original isosceles triangle and the triangle formed by the sides a, d, c. They share the angle γ, and we have just shown that the sides that form γ have the same proportion (for the original triangle, recall that $b = a + c$). So these triangles are similar, and hence $a = d$ and $\beta = \delta$.

The first conclusion means that the triangle with the sides b, d, a is also isosceles. Hence, $\alpha = \beta$, i.e., the angles α, β, and δ are equal. Also $\alpha + \delta = \gamma$, so $\alpha + \beta + \gamma + \delta = 5\beta$. But $\alpha + \beta + \gamma + \delta = 180°$, so $\beta = 36°$ and $\alpha + \delta = \gamma = 72°$.

Consider now the following drawing, where R is the point at which the fourth circle touches the smaller circle on the right.

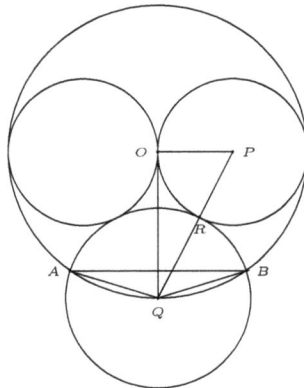

An analysis of Hirano's construction.

Assume $OQ = 1$. Then $PO = PR = \frac{1}{2}$, so by the Pythagorean theorem, $PQ = \sqrt{PO^2 + OQ^2} = \frac{\sqrt{5}}{2}$. Hence, $QB = QR = PQ - PR = \frac{\sqrt{5}-1}{2}$. This means that in the isosceles triangle OQB, we have

$$\frac{OQ}{QB} = \frac{2}{\sqrt{5}-1} = \frac{2(\sqrt{5}+1)}{(\sqrt{5}-1)(\sqrt{5}+1)} = \frac{\sqrt{5}+1}{2} = \phi.$$

From the above observation about the golden triangle, we conclude that $\angle QOB = 36°$. By symmetry $\angle QOA = 36°$, so $\angle AOB = 72°$. This shows that AB is the side of a regular pentagon inscribed in the first circle.

14   Complex roots of unity

To deal with the equations of third and fourth degrees, we first need to analyze solutions of the equation $x^3 = 1$. Clearly, one of the solutions is $x = 1$. However, this equation has also two solutions in complex numbers, namely

$$x = \frac{1}{2}(-1 + i\sqrt{3}) \text{ and } x = -\frac{1}{2}(1 + i\sqrt{3}).$$

To see it note that by formula (1) from Appendix 10

$$(-1 + i\sqrt{3})^3 = -1 + 3(i\sqrt{3}) - 3(i\sqrt{3})^2 + (i\sqrt{3})^3$$
$$= -1 + i3\sqrt{3} - 3(-3) + (-i3\sqrt{3}) = 8$$

and

$$(1 + i\sqrt{3})^3 = 1 + 3(i\sqrt{3}) + 3(i\sqrt{3})^2 + (i\sqrt{3})^3$$
$$= 1 + i3\sqrt{3} + 3(-3) + (-i3\sqrt{3}) = -8.$$

As customary, we shall denote the above 3 solutions by, respectively, ζ_1, ζ_2, and ζ_3. This means that there are 3 solutions of the equation $x^3 = r$ for an arbitrary non-negative real number r, namely

$$x = \zeta_1 \sqrt[3]{r}, \ x = \zeta_2 \sqrt[3]{r} \text{ and } x = \zeta_3 \sqrt[3]{r}.$$

We shall apply this observation in the next Appendix, where instead of r we shall use an arbitrary complex number z.

There is a more general and systematic approach, due to de Moivre, that allows us to find all n solutions of the equation $x^n = 1$ for an arbitrary $n > 1$. However, this approach is not needed to deal with the equations of third and fourth degrees, so we omit it.

15   A derivation of the solutions of third-degree equations

The general form of a third-degree equation is

$$ax^3 + bx^2 + cx + d = 0, \tag{4}$$

where a, b, c, and d are constants (called *coefficients*), such that $a \neq 0$.

To solve it, we proceed in two steps. First, we eliminate the term with x^2. Then, we transform the resulting equation into a quadratic one. Both steps are achieved by appropriate substitutions.

The first step is taken care of by substituting in the above equation x by $u - \frac{b}{3a}$, where u is a variable. After some simple calculations[7], we obtain the equation

$$au^3 + (-\frac{b^2}{3a} + c)u + (\frac{2b^3}{27a^2} - \frac{bc}{3a} + d) = 0.$$

Dividing both sides by a and introducing the constants

$$p = \frac{1}{3a}(-\frac{b^2}{3a} + c) \text{ and } q = -\frac{1}{2a}(\frac{2b^3}{27a^2} - \frac{bc}{3a} + d)$$

we are left with a simplified equation

$$u^3 + 3pu - 2q = 0, \tag{5}$$

that we reduce next to a quadratic equation.

We use the approach due to Viète, which is simpler than the method of Tartaglia published by Cardano. The trick is to substitute in equation (5) the variable u by $\frac{p}{y} - y$, where y is a variable. Simple calculations[8] show that this results in a transformation of equation (5) into the equation $-y^3 + \frac{p^3}{y^3} - 2q = 0$, or equivalently

$$y^6 + 2qy^3 - p^3 = 0, \tag{6}$$

which, after substituting y^3 by a variable z, becomes a quadratic equation

$$z^2 + 2qz - p^3 = 0$$

in z.

According to the formulas given in Appendix 11, its two solutions are

$$z = -q + \sqrt{D} \text{ and } z = -q - \sqrt{D},$$

where

$$D = q^2 + p^3.$$

Further, according to Appendix 14, the equation $y^3 = z$ has three solutions in y:

$$y = \zeta_1 \sqrt[3]{z}, \ y = \zeta_2 \sqrt[3]{z}, \text{ and } y = \zeta_3 \sqrt[3]{z},$$

where $\zeta_1, \zeta_2, \zeta_3$ are the third roots of unity.

Combining this with the above two solutions for z, we obtain six solutions in y of equation (6). But this equation has been obtained by substituting u by $\frac{p}{y} - y$ in equation (5), so we obtain six solutions in u of equation (5).

To show that these are only three solutions, we group them in the following three pairs

$$y_1 = \zeta_1 \sqrt[3]{-q + \sqrt{D}} \text{ and } y_2 = \zeta_1 \sqrt[3]{-q - \sqrt{D}},$$
$$y_1 = \zeta_2 \sqrt[3]{-q + \sqrt{D}} \text{ and } y_2 = \zeta_3 \sqrt[3]{-q - \sqrt{D}},$$
$$y_1 = \zeta_3 \sqrt[3]{-q + \sqrt{D}} \text{ and } y_2 = \zeta_2 \sqrt[3]{-q - \sqrt{D}}.$$

It is straightforward to check[9] that

$$\zeta_1 \zeta_1 = \zeta_2 \zeta_3 = \zeta_3 \zeta_2 = 1,$$

which implies, by simple calculations,[10] that for each pair of y_1 and y_2, we have $y_1 y_2 = -p$. So

$$\frac{p}{y_1} - y_1 = -y_2 - y_1 = -y_1 - y_2 = \frac{p}{y_2} - y_2.$$

Summarizing, we have just three solutions of equation (5). They are of the form

$$u = \frac{p}{y} - y,$$

where

$$y = \zeta_1 \sqrt[3]{-q + \sqrt{D}} \text{ or } y = \zeta_2 \sqrt[3]{-q + \sqrt{D}} \text{ or } y = \zeta_3 \sqrt[3]{-q + \sqrt{D}} \quad (7)$$

with p, q, and D defined earlier.

Finally, recall that p and q were expressed in terms of the coefficients of the original equation (4) and that equation (5) was obtained by substituting x by $u - \frac{b}{3a}$ in equation (4). So each solution u of equation (5) yields a solution $u - \frac{b}{3a}$ of the original equation (4).

To summarize, each solution of the original equation (4) is of the form $\frac{p}{y} - y - \frac{b}{3a}$, with y being one of the three values defined in equation (7).

16   A derivation of the solutions of fourth-degree equations

The general form of a fourth-degree equation is

$$ax^4 + bx^3 + cx^2 + dx + e = 0, \quad (8)$$

where a, b, c, d, and e are coefficients such that $a \neq 0$.

To solve it, as in the previous Appendix, we proceed in two steps. First, we eliminate the term with x^3. Then, we transform the resulting equation into an equation of the third degree.

The first step is taken care of by substituting in the above equation x by $y - \frac{b}{4a}$, where y is a variable. After some simple calculations,[11] we obtain the equation

$$ay^4 - \frac{3b^2 - 8ac}{8a}y^2 + \frac{b^3 - 4abc + 8a^2 d}{8a^2}y$$
$$+ \frac{-3b^4 + 16ab^2 c - 64a^2 bd + 256a^3 e}{256a^3} = 0.$$

Dividing both sides by a and introducing the constants

$$p = -\frac{3b^2 - 8ac}{8a^2},$$
$$q = \frac{b^3 - 4abc + 8a^2 d}{8a^3},$$
$$r = \frac{-3b^4 + 16ab^2 c - 64a^2 bd + 256a^3 e}{256a^4}$$

we are left with a simplified equation

$$y^4 + py^2 + qy + r = 0, \tag{9}$$

that we solve next.

We present an approach due to Descartes that is simpler than the method originally proposed by Ferrari. We want to represent the left-hand side as the product

$$y^4 + py^2 + qy + r = (y^2 + sy + t)(y^2 + uy + v) \tag{10}$$

for some still to be determined s, t, u, and v. By multiplying and comparing the coefficients of the powers of y, we obtain the equations

$$0 = s + u,$$
$$p = v + su + t,$$
$$q = tu + vs,$$
$$r = tv,$$

which after some simple manipulations,[12] yield the equation

$$u^6 + 2pu^4 + (p^2 - 4r)u^2 - q^2 = 0$$

in u. After substituting u^2 by a variable z, we obtain a third-degree equation

$$z^3 + 2pz^2 + (p^2 - 4r)z - q^2 = 0$$

in z. Using the formulas given in the previous appendix, we obtain three solutions for z, i.e., three solutions for u^2. This allows us to express s, t, and v in terms of u^2 and the previously introduced constants p and q and hence solve both equations $y^2 + sy + t = 0$ and $y^2 + uy + v = 0$ introduced in equation (10).[13] By equation (10), the resulting solutions are also solutions of equation (9).

Finally, as in the previous appendix, each solution y of (9) yields a solution $y - \frac{b}{4a}$ of the original equation (8).

17   A walled town problem

This problem was introduced in Chapter 4. Let us recall its formulation.

> There is a round, walled town of unknown diameter with four
> gates. A tree lies 3 li north of the northern gate. If one walks 9
> li eastward from the southern gate, the tree becomes just visible.
> Find the diameter of the town.

This problem is discussed in detail in a book by C. Smoryński, in which three
solutions to this problem are given.[14] Here is another solution that makes use
of a trigonometric identity concerned with the tangent function tan.

It is more informative to reason about arbitrary values a and b instead of 9
and 3. So the problem is to compute the diameter x of the circle, given a and
b. Add an additional radius of the circle to the original drawing, as shown in
the figure below.

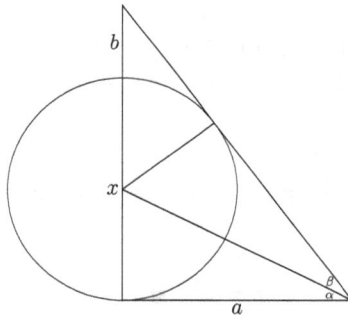

The round walled town problem revisited.

Consider right-angled triangles with the angles α and β. They share one
side, and in each of them, another side equals to the radius of the circle. So,
by the Pythagorean theorem, the third sides are equal as well.

Hence, these triangles are congruent, and, consequently, $\alpha = \beta$. We have
$\tan(2\alpha) = \frac{x+b}{a}$ and $\tan(\alpha) = \frac{\frac{x}{2}}{a}$, so

$$\tan(2\alpha) = 2\tan(\alpha) + \frac{b}{a}.$$

We now use the trigonometric identity

$$\tan(2\alpha) = \frac{2\tan(\alpha)}{1 - \tan^2(\alpha)}.$$

The last two equalities lead to the equation

$$2y + \frac{b}{a} = \frac{2y}{1 - y^2},$$

where $y = \tan(\alpha)$.

Multiplying both sides by $a(1 - y^2)$ and simplifying, we get a third-degree equation in y

$$2ay^3 + by^2 - b = 0.$$

But $y = \frac{x}{2}$, so by substituting, we get after simplifying the following third-degree equation in x:

$$x^3 + bx^2 - 4a^2b = 0.$$

For $a = 9$ and $b = 3$, we get three solutions using *WolframAlpha*—two in complex numbers and one in real numbers, $x = 9$, which is the sought answer.

18 How logarithms work and why they are useful

First, let us recall the logarithm function (written as log). We limit ourselves to base 10; that is, we only consider powers of 10. (To avoid confusion, one sometimes writes $\log_{10}$.) If $x = 10^y$, then by definition, $\log(x) = y$. In other words, the logarithm of a real number x is the power to which 10 needs to be raised to get x.

Hence, $\log(x_1) = y_1$ and $\log(x_2) = y_2$ means that $x_1 = 10^{y_1}$ and $x_2 = 10^{y_2}$. This implies that

$$x_1 \cdot x_2 = 10^{y_1} \cdot 10^{y_2} = 10^{y_1 + y_2},$$

i.e., $\log(x_1 \cdot x_2) = y_1 + y_2$.

In other words, using different variables

$$\log(x \cdot y) = \log(x) + \log(y).$$

This is the key property of the logarithm function. Namely, it shows that in order to compute the product of x and y, it suffices to compute the sum $\log(x) + \log(y)$ of their logarithms, and then look up a table to find the number whose logarithm is $\log(x) + \log(y)$. Similarly, to compute the division of x by y, one can use the law

$$\log\left(\frac{x}{y}\right) = \log(x) - \log(y)$$

and look up the table to find the number whose logarithm is $\log(x) - \log(y)$.

Using base 10 is useful because we use the decimal system. Note that for an integer k, we have

$$\log(10^k x) = \log(10^k) + \log(x) = k + \log(x).$$

So one can just produce the table of logarithms of natural numbers in a given range, say $1..10^6$, and then use it to determine the logarithms of numbers of a smaller (using negative k) or larger magnitude (using positive k).

As an example, let us compute the approximate value of the product of π and e, by rounding both numbers to three decimal places (3.141 and 2.718), using addition and the table of logarithms just mentioned (we use *WolframAlpha* instead).

We have

$$\log(3.141) = \log(10^{-3} \cdot 3141) = -3 + \log(3141) \approx -3 + 3.4970 = 0.4970$$

and

$$\log(2.718) = \log(10^{-3} \cdot 2718) = -3 + \log(2718) \approx -3 + 3.4342 = 0.4342.$$

So $\log(3.141 \cdot 2.718) \approx 0.4970 + 0.4342 = 0.9312$. Logarithms of natural numbers from the range $1..10^6$ lie between 0 and 6. To get a better precision, consider the value 5.9312 for which we have $\log(853493) \approx 5.9312$. Thus

$$\log(8.53493) = \log(10^{-5} \cdot 853493) = -5 + \log(853493) \approx 0.9312.$$

We conclude that $\pi \cdot e \approx 8.53493$, which is correct to two decimal places.

19 Euler's formula

Recall that a graph is a set of vertices, some of which are connected by edges. Euler's formula deals with graphs. In the next two appendices, we shall use it to solve the gas, water, and electricity puzzle from Chapter 6 and to prove that there are exactly five Platonic solids.

This formula concerns an arbitrary connected graph (so, a graph in which each vertex can be reached, perhaps indirectly, from any other one), which is planar (it can be drawn on a plane in such a way that no two edges cross). It states that for such graphs

$$V - E + F = 2,$$

where V is the number of vertices, E is the number of edges, and F is the number of faces (regions) in a given graph. (Why the word 'face' is used will become clear in Appendix 21.)

To illustrate it, reconsider the drawing on page 82 providing an incomplete solution to the gas, water, and electricity puzzle.

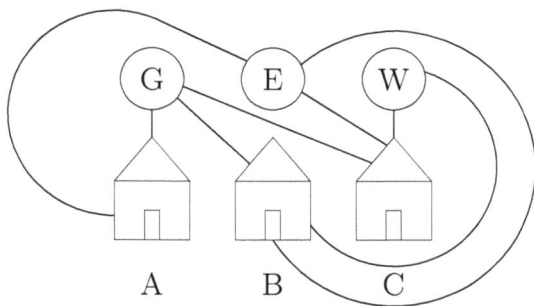

The gas, water, and electricity puzzle.

Here we have

- $V = 6$, namely, A, B, C, E, G, W,
- $E = 8$, namely, $AE, AG, BE, BG, BW, CE, CG, CW$,
- $F = 4$, namely, there is the region outside the graph and three regions to which the houses A, B, and C belong, respectively, so the regions formed by the cycles $AECGA$, $BECWB$, and $CGBWC$.

Note that $6 - 8 + 4 = 2$.

Proof of Euler's formula. Consider a connected planar graph. It is useful to envisage it as a field covered with water in which some edges of the graph form dams. Each face is then a separate water region, and their number equals F.

We are interested in the value of the expression $V - E + F$. If the graph has a cycle, we then remove an edge from it. Then, V does not change, and E drops by 1. Also, F drops by 1 because by removing an edge from a cycle, we 'break' the dam between two faces. So $V - E + F$ does not change. Eventually, we reach a graph without cycles, called a *tree*. Then there is only one face, and no edge forms a dam. A vertex in a tree connected to only one vertex is called a *leaf*.

From now on, we pick an edge that reaches a leaf of the tree and remove it, together with the leaf. Each time, both V and E drop by 1, while F does not change (it remains 1). So again, $V - E + F$ does not change. Eventually, we reach a single point. At this stage, $V = 1$, $E = 0$, and $F = 1$, and we conclude that $V - E + F = 2$.

20   A solution to the gas, water, and electricity puzzle

Using Euler's formula, we now solve the gas, water, and electricity puzzle from page 82. Suppose it does have a solution. This means that there is a planar graph with the vertices A, B, C, W, G, E and the edges

$$AE, AG, AW, BE, BG, BW, CE, CG, CW.$$

First, note that the resulting graph is connected. Indeed, any house can be reached from another through any utility, and any utility can be reached from another through any house.

In the assumed solution graph, we have $V = 6$ and $E = 9$, so by Euler's formula, $F = 2 + E - V = 5$.

Let us now estimate how many edges are needed to form each face. By definition, there are no faces formed by 2 edges. Further, there are no faces formed by 3 edges either. Indeed, otherwise one of these edges would connect two houses or two utilities, while, by definition, each edge connects a house and a utility. This means that each face is formed by at least 4 edges.

Thus the faces of the solution graph use at least $4 \cdot 5 = 20$ edges. But there are only 9 edges. This means that some edge was used in at least three faces, which is impossible, as each edge can belong to at most two faces.

21   A proof that there are exactly five Platonic solids

The proof applies Euler's formula to regular and convex polyhedra. They were defined in Chapter 2 as three-dimensional objects enclosed by identical flat faces, the vertices of which lie on a sphere.

We first transform a given regular and convex polyhedron to a connected planar graph as follows. Imagine that all the edges are made from rubber. Remove one face of the polyhedron, stretch the edges of this face outwards, and subsequently flatten the resulting object so that all the other edges end up inside the polygon formed by the stretched edges of the removed face. The following diagram illustrates this operation applied to a cube in which the edges of the face $EFGH$ are stretched outwards.

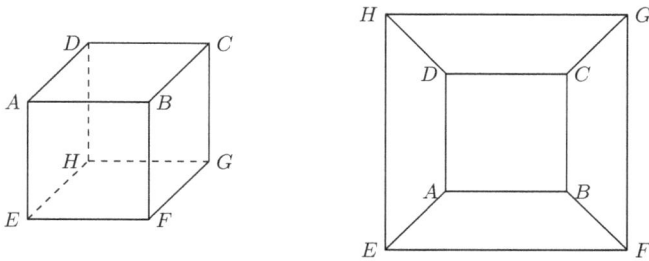

A cube and its flattened version.

Since the resulting graph is connected and planar, Euler's formula $V - E + F = 2$ holds for it. But this graph has the same number of vertices and edges as the original polyhedron. Moreover, the number of faces is also the same since the removed face now corresponds to the outside region, and all other faces remain (though deformed). So Euler's formula, $V - E + F = 2$, also holds for the original regular and convex polyhedron on which we now focus.

Let v be the number of edges leaving each vertex. This means that the total number of edges E equals $\frac{vV}{2}$, where we divide by 2 since each edge connects two vertices and hence is counted in vV twice.

Let e be the number of edges of each face. This means that the total number of edges E also equals $\frac{eF}{2}$, where we divide by 2 since each edge belongs to two faces.

So we have $V = \frac{2E}{v}$ and $F = \frac{2E}{e}$. By substituting it in $V - E + F = 2$, we get $\frac{2E}{v} - E + \frac{2E}{e} = 2$, i.e., $E(\frac{2}{v} - 1 + \frac{2}{e}) = 2$. Hence, $\frac{2}{v} - 1 + \frac{2}{e} > 0$, or equivalently

$$\frac{1}{v} + \frac{1}{e} > \frac{1}{2}. \tag{11}$$

Each face has at least three edges, so $e \geq 3$, or equivalently $\frac{1}{e} \leq \frac{1}{3}$. Hence,

$$\frac{1}{v} > \frac{1}{2} - \frac{1}{e} \geq \frac{1}{2} - \frac{1}{3} = \frac{1}{6},$$

or equivalently $v < 6$. Further, at least three edges are leaving each vertex, so $v \geq 3$. By the analogous argument with v and e interchanged we get $e < 6$.

We conclude that for both e and v, the only possible values are $3, 4,$ or 5. Out of the nine combinations $(3,3), (3,4), (3,5), (4,3), (4,4), (4,5), (5,3), (5,4), (5,5)$ for e and v only

$$(3,3), (3,4), (3,5), (4,3), (5,3)$$

satisfy condition (11). Each such pair of values for e and v uniquely determines one type of a polyhedron. So, we conclude that there are only five possible regular convex polyhedra. These are the ones we already encountered in Chapter 2.

22   Condorcet paradox and Borda's count

Condorcet proposed a voting method based on the idea of comparing candidates pairwise. Consider an election in which each voter ranks all the candidates. Let A and B be two candidates. We say then that A *beats* B (in a direct election) if a majority of voters ranked A higher than B. Then, a candidate is called a *Condorcet winner* if he/she beats every other candidate. Condorcet then observed the following complication, now called the *Condorcet paradox*.

To illustrate it, suppose there are three candidates A, B, and C, and 60 voters whose preferences among candidates (expressed using the $>$ symbol) are respectively:

- 23 voters: $A > B > C$,
- 2 voters: $B > A > C$,
- 17 voters: $B > C > A$,
- 10 voters: $C > A > B$,
- 8 voters: $C > B > A$.

Note that

- a majority of voters (33) prefers A to B,
- a majority of voters (42) prefers B to C,
- a majority of voters (35) prefers C to A,

so there is no Condorcet winner. This means that it is difficult to argue who should be a winner in this election by using this voting method.

Note also that candidate A was put most often on the top of a ballot, so according to a natural *plurality rule*, he should be declared a winner.

Recall the voting method proposed by Borda. Assuming there are n candidates, each voter gives n points to his/her most preferred candidate, $n-1$ points to the second most preferred candidate, and so on. The winner is the candidate who obtained the largest number of points.

Condorcet criticized the Borda count using the following example. Suppose there are three candidates, A, B, and C, and 81 voters whose preferences among candidates are respectively:

- 30 voters: $A > B > C$,
- 1 voter: $A > C > B$,
- 29 voters: $B > A > C$,
- 10 voters: $B > C > A$,
- 10 voters: $C > A > B$,

- 1 voter: $C > B > A$.

Then

- a majority of voters (41) prefers A to B,
- a majority of voters (60) prefers A to C.

So A is the Condorcet winner. However, according to the Borda count

- A gets $31 \cdot 3 + 39 \cdot 2 + 11 \cdot 1 = 182$ points,
- B gets $39 \cdot 3 + 31 \cdot 2 + 11 \cdot 1 = 190$ points,
- C gets $11 \cdot 3 + 11 \cdot 2 + 59 \cdot 1 = 114$ points,

so B is the winner.

Finally, the following simple example shows that the Borda count does not solve the Condorcet paradox. Suppose there are three candidates, A, B, and C, and three voters, 1, 2, and 3, with the following preferences among candidates:

- voter 1: $A > B > C$,
- voter 2: $B > C > A$,
- voter 3: $C > A > B$.

Then

- a majority (voters 1 and 3) prefer A over B,
- a majority (voters 1 and 2) prefer B over C,
- a majority (voters 2 and 3) prefer C over A.

So there is no Condorcet winner. Further, according to the Borda count, each candidate gets 6 points, so this method does not yield a winner either.

23 Jefferson's method

We explain Jefferson's method of apportionment, using the original data to which it was applied.[15]

In 1792, the population of the United States was 3,615,920 and there were 105 seats in Congress that needed to be divided among 15 states. So, the number of citizens per seat, after rounding off, is

$$D = \frac{3,615,920}{105} \approx 34,437.3333.$$

To obtain the number of seats per state, we use the data from the table below and first divide the population of each state by D. For example, for

Virginia (VA), we get $630,560/34,437.3333 \approx 18.3103$. The number of seats has to be a natural number, so we round this number down to 18. Proceeding this way for each state, we distribute the seats among all states, but there is a problem because the total is 97 instead of 105.

To solve it, we decrease D by an appropriate number d so that the above procedure yields 105 seats. The correct value turns out to be $d = 1279$.[16] The following table summarizes this discussion, where each *quota* is the result of the division of each state population by $D - d$. For example, in the case of Virginia, the quota equals

$630,560/(D - d) = 630,560/(34,437.3333 - 1279) \approx 19.0166373461.$

state	population	first outcome	quota	apportionment
VA	630,560	18	19.0166373461	19
MA	475,327	13	14.3350691128	14
PA	432,879	12	13.0549082684	13
NC	353,523	10	10.6616637346	10
NY	331,589	9	10.0001708972	10
MD	278,514	8	8.3995174667	8
CT	236,841	6	7.142729329	7
SC	206,236	5	6.2197336014	6
NJ	179,570	5	5.4155315406	5
NH	141,822	4	4.277114853	4
VT	85,533	2	2.5795325459	2
GA	70,835	2	2.1362653933	2
KY	68,705	1	2.0720281478	2
RI	68,446	1	2.06421714	2
DE	55,540	1	1.674993717	1
Total:	3,615,920	97		Total: 105

The method of D'Hondt avoids an ad hoc approach to finding d by consecutively assigning the seats to the currently largest party (here, state) after the allocated seat has been accounted for, and the number of votes (here, population) has been adjusted.

24   The birthday paradox

Recall the question posed in Chapter 6:

> What is the probability that in a random group of 23 people, two of them share the same birthday?

To answer this, we first solve the dual problem: what is the probability that among randomly chosen 23 people, everybody has a unique birthday? By a *birthday sequence*, we mean a sequence of 23 birth dates (defined as a day-month pair; we ignore leap years).

We can choose each birth date arbitrarily, so in 365 ways. So the total number of birthday sequences is 365^{23}. In contrast, the total number of birthday sequences in which no two birthdays are the same is $365 \cdot 364 \cdot \ldots \cdot 343$. Indeed, we can choose the first birth date arbitrarily, so in 365 ways. After that, we can choose the second one only in 364 ways; subsequently, the third one in 363 ways, and so on. So the probability that in a birthday sequence, no two birth dates are the same is

$$\frac{365 \cdot 364 \cdot \ldots \cdot 343}{365^{23}},$$

which is approximately 0.4927. Hence, the answer to the original question is approximately 0.5073.

25 A minimal introduction to predicate logic

To clarify the idea of predicate logic, let us first formalize Fermat's Last Theorem discussed in Chapter 8. Recall that it states that for $n > 2$, the equation

$$x^n + y^n = z^n$$

has no solutions in natural numbers. It can be formalized as the following predicate logic formula about natural numbers

$$\forall x \, \forall y \, \forall z \, \forall n \, (n > 2 \rightarrow x^n + y^n \neq z^n),$$

where $\forall x$ stands for 'for all x' and $\rightarrow$ stands for the implication.

For another example, let us return to the proof that there are infinitely many prime numbers, given in Appendix 5. We noted there that it is sufficient to establish the following statement:

For every natural number, there is a prime number larger than it.   (12)

We now express it as a formula in predicate logic about natural numbers. First, we introduce the abbreviation *pdiv*(x, y) for the formula

$$x > 1 \wedge \exists z \, (x \cdot z = y),$$

where $\wedge$ stands for the conjunction ('and') and $\exists z$ stands for 'there exists z'. So *pdiv*(x, y) holds if x is a proper divisor of y. Next, we write *prime*(y) for the formula

$$\forall x \, (\textit{pdiv}(x, y) \rightarrow x = y).$$

So *prime*(y) holds if y is a prime number.

Then, the formula

$$\forall x\, \exists y\, (x < y \wedge prime(y)),$$

expresses (12). The reader should compare it with the formulation in Frege's notation presented on page 105.

Finally, to see how more complex properties can be expressed in predicate logic, consider the statement

The smallest proper divisor of $n! + 1$ is a prime number larger than n

established in Appendix 5. We can express it as the following formula in predicate logic:

$$\forall x\, (pdiv(x, n! + 1) \wedge [\neg \exists y\, (pdiv(y, n! + 1) \wedge y < x)] \rightarrow [prime(x) \wedge x > n]),$$

where $\neg$ stands for the negation.

In contrast, one can show that the following statement, called the *Geach–Kaplan* sentence,

Some critics admire only one another

cannot be formalized in predicate logic.

26   Kolmogorov's axioms of probability

To introduce Kolmogorov's approach to probability, we need a couple of concepts. Assume some experiment with a number of possible outcomes. In what follows, we limit ourselves to experiments with a finite set of outcomes, for example, throwing a dice.

An *event* is a set of outcomes in this experiment. An *elementary event* is an event with a single outcome. The *sure event*, denoted by S, is the event consisting of all outcomes. We say that two events A and B are *mutually exclusive* if they cannot both occur at the same time. Since each event is a set of outcomes, one uses for it the set-theoretic notation and writes it as $A \cap B = \emptyset$.

For example, throwing a dice yields 6 possible outcomes. Each of them corresponds to one elementary event. An example of an event is 'the outcome of throwing the dice is bigger than 4', which consists of two possible outcomes of the throw: 5 or 6.

Given an event A, we denote its *probability* by $P(A)$. Kolmogorov's axioms of probability are as follows.

Axiom 1 The probability of each event is non-negative, i.e., for each event A

$$P(A) \geq 0.$$

Axiom 2 The probability of the sure event is 1, i.e.,

$$P(S) = 1.$$

Axiom 3 For mutually exclusive events A and B

$$P(A \cup B) = P(A) + P(B),$$

where $A \cup B$ denotes the event that one of the events A and B took place. This axiom naturally generalizes to an arbitrary number of mutually exclusive events.

To illustrate these axioms, let us return to the experiment of throwing a dice. The sure event S consists of all possible outcomes. By Axiom 2, we have $P(S) = 1$.

Assume now that the dice is fair, which means that the probability of each of the six outcomes is the same. Denote by E_i the event that the outcome of the dice throw is i. So, for each $i \in \{1, 2, 3, 4, 5, 6\}$, we have $P(E_i) = \frac{1}{6}$. (We use here and below the set-theoretic notation, according to which $\{1, 2, 3, 4, 5, 6\}$ is a set consisting of the listed elements and '$i \in \{1, 2, 3, 4, 5, 6\}$' means that i is an element of this set.)

The events of throwing 5 and throwing 6, so E_5 and E_6, are independent. Denote the earlier mentioned event 'the outcome of throwing the dice is bigger than 4' by E. Then by Axiom 3

$$P(E) = P(E_5 \cup E_6) = \frac{1}{6} + \frac{1}{6} = \frac{1}{3}.$$

When discussing Bayes' theorem in the next Appendix, we shall refer to one more notion. Assuming that $P(B) > 0$ and that event B took place, the probability that event A takes place is denoted by $P(A \mid B)$. The expression $A \mid B$ should be read as 'A given B'.

$P(A \mid B)$ is called the *conditional probability*. It can be computed using the formula

$$P(A \mid B) = \frac{P(A \cap B)}{P(B)}, \tag{13}$$

where $A \cap B$ is an event stating that both events A and B took place. (Recall that we assumed that $P(B) > 0$.)

As an example, suppose we threw two dice. Consider the following two events:

- A: one outcome is 4,
- B: the outcomes of two throws are different.

Then, given that the outcomes of two throws are different, the probability that one outcome is 4 is $P(A \mid B)$. Here, $A \cap B$ is the event that the outcomes of the two throws are different *and* one of them is 4. We have $P(A \cap B) = \frac{10}{36}$, since out of 36 outcomes, exactly 10 are 'good', namely $(4, x)$ and $(x, 4)$, where $x \in \{1, 2, 3, 5, 6\}$.

Further, $P(B) = \frac{30}{36}$ because out of 36 outcomes, exactly 6 are 'bad', namely, (x, x), where $x \in \{1, 2, 3, 4, 5, 6\}$, so 30 outcomes are 'good'. So by formula (13)

$$P(A \mid B) = \frac{\frac{10}{36}}{\frac{30}{36}} = \frac{1}{3}.$$

27 Bayes' theorem

Bayes' theorem is concerned with conditional probabilities. Note that $A \cap B$ and $B \cap A$ are the same events, so $P(A \cap B) = P(B \cap A)$. In contrast, the conditional probabilities $P(A \mid B)$ and $P(B \mid A)$ in general differ. How one of them can be computed from the other is the subject of Bayes' theorem. It states that

$$P(A \mid B) = P(B \mid A) \cdot \frac{P(A)}{P(B)}$$

for any two events A and B with $P(B) > 0$.

The proof is very simple. By formula (13) for the conditional probability, we have both

$$P(A \mid B) \cdot P(B) = P(A \cap B)$$

and (by exchanging A and B)

$$P(B \mid A) \cdot P(A) = P(B \cap A).$$

But $A \cap B$ and $B \cap A$ are the same events, so

$$P(A \mid B) \cdot P(B) = P(B \mid A) \cdot P(A),$$

from which the result follows.

To illustrate how Bayes' theorem is used, let us return to the example considered at the end of the previous Appendix. We now compute the probability that the outcomes of two dice throws are different (event B), given that one outcome is 4 (event A). So we want to compute $P(B \mid A)$.

We have $P(A) = \frac{11}{36}$, because out of 36 outcomes, exactly 25 are 'bad', namely, (x, y), where both $x \in \{1, 2, 3, 5, 6\}$ and $y \in \{1, 2, 3, 5, 6\}$, so 11 outcomes are 'good'. Also, recall that $P(B) = \frac{30}{36}$.

We now use Bayes' theorem with A and B interchanged. We already checked in the previous Appendix that $P(A \mid B) = \frac{1}{3}$. So

$$P(B \mid A) = P(A \mid B)\frac{P(B)}{P(A)} = \frac{1}{3} \cdot \frac{\frac{30}{36}}{\frac{11}{36}} = \frac{10}{11}.$$

28 The Monty Hall problem: an informal analysis

This problem was introduced in Chapter 6. Let us recall its formulation.

> Imagine a stage with three doors. Behind one of them, there is a prize, say a car, and behind each of the other two, a goat. The participant is asked to pick a door. Say he chooses door 1. Subsequently, the show host opens another door, say door 2, revealing a goat. The participant is offered to reconsider his choice. The question is: should he switch to door 3?

We have three events that appear in succession:

- a car is hidden behind a door,
- the participant selects a door,
- the host selects a door.

As there are 3 doors and 3 events, it seems that we have to do with $3^3 = 27$ possible situations. However, the additional assumption is that the host selects a door, which is different from the one selected by the participant and from the one behind which there is a car. This limits the number of legal possibilities to 18. Here is the listing of all 27 situations where a triple (a, b, c) means that the car is hidden behind door a, the participant selected door b and the host selected door c, and where the illegal combinations are crossed out:

~~(1,1,1)~~, (1,1,2), (1,1,3), ~~(1,2,1)~~, ~~(1,2,2)~~, (1,2,3), ~~(1,3,1)~~, (1,3,2), ~~(1,3,3)~~,
~~(2,1,1)~~, ~~(2,1,2)~~, (2,1,3), (2,2,1), ~~(2,2,2)~~, (2,2,3), (2,3,1), ~~(2,3,2)~~, ~~(2,3,3)~~,
~~(3,1,1)~~, (3,1,2), ~~(3,1,3)~~, (3,2,1), ~~(3,2,2)~~, ~~(3,2,3)~~, (3,3,1), (3,3,2), ~~(3,3,3)~~.

For example, situation (1,2,1) is illegal because it states that the car is hidden behind door 1 and the host selected door 1 as well. This leaves us with 12 legal combinations. In six of them, namely,

(1,1,2), (1,1,3), (2,2,1), (2,2,3), (3,3,1), (3,3,2),

the car is behind the door selected by the participant, while in the remaining 6, so

$$(1,2,3), (1,3,2), (2,1,3), (2,3,1), (3,1,2), (3,2,1),$$

the car is behind the door not selected by the participant. So it seems that the chance that the car is behind the other door is $\frac{1}{2}$, which indicates that the participant does not gain anything by switching.

However, this analysis is incorrect as it does not take into account the fact that the host selected the door *after* the participant did. As a result, the 12 legal combinations are not equally likely.

Indeed, the probability of each of the six situations in which the participant should not switch is only $\frac{1}{18}$ and not $\frac{1}{12}$. Consider, for example, situation (1,1,2). The probability that a situation starts with the pair (1,1) is $\frac{1}{3} \cdot \frac{1}{3} = \frac{1}{9}$ because hiding the car and the selection of the door by the participant are independent events. On the other hand, once the initial pair (1,1) is selected, the probability that the host selects door 2 is $\frac{1}{2}$, since his equally likely options are only 2 and 3. So, the overall probability that the situation (1,1,2) takes place is $\frac{1}{9} \cdot \frac{1}{2} = \frac{1}{18}$.

The reasoning for the other five situations in which the participant should not switch is the same. So, the overall probability that a situation arises in which the car is behind the door selected by the participant is only $6 \cdot \frac{1}{18} = \frac{1}{3}$. This explains why the participant should switch. The diagram below lists all 12 legal combinations together with the resulting probabilities.

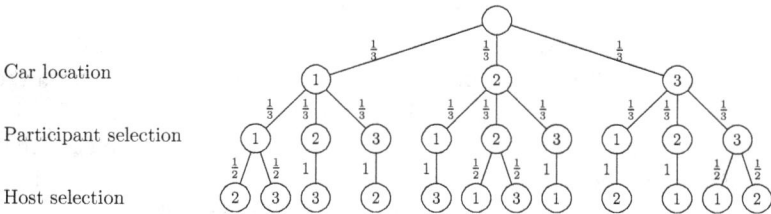

The Monty Hall problem.

29 The Monty Hall problem and Bayes' theorem

We now provide a more formal analysis of the problem based on Bayes' theorem. Let us return to the problem description. The car was hidden behind a door, then the participant chose door 1, and, subsequently the host opened door 2. We define the relevant events:

- $A = 1$: the car is behind door 1,
- $A = 2$: the car is behind door 2,
- $A = 3$: the car is behind door 3,
- B: the host opened door 2.

We want to determine the probability that the car is behind door 1, given that the host opened door 2, so $P(A = 1 \mid B)$.

By Bayes' theorem

$$P(A = 1 \mid B) = \frac{P(B \mid A = 1) \cdot P(A = 1)}{P(B)}. \tag{14}$$

First, note that

- $P(B \mid A = 1) = \frac{1}{2}$, because, assuming that event $A = 1$ took place, the host can choose door 2 or 3 with equal probability,
- $P(B \mid A = 2) = 0$, because the host never opens the door with the car behind it (in this case, door 2),
- $P(B \mid A = 3) = 1$, because the host always opens the door different than door 1 and he never opens the door with the car behind it (so in this case, door 3).

Further, we have $P(A = 1) = P(A = 2) = P(A = 3) = \frac{1}{3}$.

So, by equation (14) to compute $P(A = 1 \mid B)$, it remains to compute $P(B)$. To do this, note that we can split event B into three mutually exclusive events $B \cap (A = 1)$, $B \cap (A = 2)$, and $B \cap (A = 3)$, that is

$$B = (B \cap (A = 1)) \cup (B \cap (A = 2)) \cup (B \cap (A = 3)).$$

(Here, we generalize the notion of mutually exclusive events to more events than just two.) We then have by Kolmogorov's Axiom 3 applied to three mutually exclusive events

$$P(B) = P(B \cap (A = 1)) + P(B \cap (A = 2)) + P(B \cap (A = 3)). \tag{15}$$

By formula (13) about the conditional probability and the just made observations about $P(B \mid A = i)$ and $P(A = i)$ for $i \in \{1, 2, 3\}$, we have

- $P(B \cap (A = 1)) = P(B \mid A = 1) \cdot P(A = 1) = \frac{1}{2} \cdot \frac{1}{3} = \frac{1}{6}$,
- $P(B \cap (A = 2)) = P(B \mid A = 2) \cdot P(A = 2) = 0 \cdot \frac{1}{3} = 0$,
- $P(B \cap (A = 3)) = P(B \mid A = 3) \cdot P(A = 3) = 1 \cdot \frac{1}{3} = \frac{1}{3}$.

So, by equation (15)

$$P(B) = \frac{1}{6} + 0 + \frac{1}{3} = \frac{1}{2}.$$

So, plugging into equation (14) the values of the relevant probabilities computed above, we get

$$P(A = 1 \mid B) = \frac{P(B \mid A = 1) \cdot P(A = 1)}{P(B)} = \frac{\frac{1}{2} \cdot \frac{1}{3}}{\frac{1}{2}} = \frac{1}{3}.$$

So, given that the host opened door 2 (event B), the chance that the car is behind door 1 (event $A = 1$) is $\frac{1}{3}$, and hence, the chance that the car is behind door 3 is $\frac{2}{3}$. The conclusion is that the participant should switch.

30   Steinhaus–Trybuła paradox

Consider bus lines X, Y, and Z, each of which can arrive at 1, 2, 3, 4, or 5 (o'clock). Assume that

- bus X arrives at 2 with probability $\frac{2}{3}$ and at 5 with probability $\frac{1}{3}$,
- bus Y always arrives at 3,
- bus Z arrives at 1 with probability $\frac{1}{3}$ and at 4 with probability $\frac{2}{3}$.

The following table summarizes these assumptions:

Bus	Arrival time	Probability
X	2	$\frac{2}{3}$
X	5	$\frac{1}{3}$
Y	3	1
Z	1	$\frac{1}{3}$
Z	4	$\frac{2}{3}$

Denote by $X = 2$ the event that bus (of line) X arrives at 2, by $X < Y$ the event that bus X arrives earlier than bus Y, and similarly with other events. So $P(Y < Z)$ is the probability that bus Y arrives earlier than bus Z, etc. We then have

- $P(X = 2) = \frac{2}{3}$, $P(X = 5) = \frac{1}{3}$,
- $P(Y = 3) = 1$,
- $P(Z = 1) = \frac{1}{3}$, $P(Z = 4) = \frac{2}{3}$,
- $P(X < Y) = P(X = 2) = \frac{2}{3} > \frac{1}{2}$,
- $P(Y < Z) = P(Z = 4) = \frac{2}{3} > \frac{1}{2}$,
- $P(Z < X) = P(Z = 1) + P(Z = 4) \cdot P(X = 5) = \frac{1}{3} + \frac{2}{3} \cdot \frac{1}{3} = \frac{5}{9} > \frac{1}{2}$,

where in the last calculations we applied Kolmogorov's Axiom 3 from Appendix 26 since the events that bus Z arrives at 4 ($Z = 4$) and bus X arrives at 5 ($X = 5$) are mutually exclusive. So all three probabilities are bigger than 50%, which means that we cannot determine which bus will most likely arrive first.

31 Russell's paradox

This paradox can be explained as follows. Call a set *proper* if it is not a member of itself. Now take set X of all proper sets. Let us ask the question: is set X proper?

Suppose the answer is 'yes'. Then, by the definition of a proper set, X is not a member of itself. So, by its definition, set X is not proper.

Suppose next that the answer is 'no'. Then, by the definition of a proper set, X is a member of itself. So, by its definition, set X is proper.

We conclude that set X is neither proper nor improper.

Expanding the notation introduced in Appendix 26, we can write that set Y is proper if $Y \notin Y$, where $\notin$ should be read as 'not an element of', and define set X succinctly as

$$X = \{Y \mid Y \notin Y\}.$$

Here, '$Y \mid$' should be read as: 'Y such that'. So X is the set of all sets Y such that $Y \notin Y$.

Then we can summarize the above reasoning about set X more formally as follows. If $X \notin X$ (X is proper), then $X \in X$ (X is not proper). And if $X \in X$ (X is not proper), then $X \notin X$ (X is proper). So, the statement $X \notin X$ is equivalent to $X \in X$, which is a contradiction.

This, shows that a naive concept of a set can lead to a contradiction.

32 Prisoner's dilemma

The prisoner's dilemma was mentioned on page 125. It is an example of a two-player strategic game. In such games, each player simultaneously chooses a strategy (an action), which results in a payoff to each player. So a player's payoff depends not only on his choice but also on the choices of his opponents. The prisoner's dilemma game illustrates the problem that, in some situations, the best outcome for both players will not be reached if each of them is concerned exclusively with maximizing his own profit.

We assume that in this game, each player has two strategies, to cooperate (denoted by C) or not (usually denoted by D, for *defect*). To define the game, we need to define the payoffs for both players for each combination of their strategies. This can be conveniently done by means of the following matrix:

	C	D
C	2, 2	0, 3
D	3, 0	1, 1

So, there are two players, the 'row' player and the 'column' player. If their choices are, respectively, C and D, then their respective payoffs are 0 and 3, and similarly for the remaining three combinations. Here is a particularly simple interpretation of these payoffs where the players report their decisions to somebody in charge of payments.[17]

> Each player decides whether he will receive 1 euro (strategy D) or the other player will receive 2 euros (strategy C). The decisions are simultaneous and independent.

For example, the above outcome $(0, 3)$ is reached when the first player decides that the other player receives 2 euros and the second player decides to receive 1 euro.

Assuming that each player aims to maximize his profit, we can determine the outcome of the game by noting that for the first player, it is always better to choose D, no matter what the second player chooses. Indeed, his payoffs are then, respectively, 3 (when the second player chooses C) and 1 (when the second player chooses D), instead of, respectively, 2 and 0. By a symmetric argument, it is also always better for the second player to choose D.

We conclude that both players choose D and thus each receives 1. However, if they both chose C, each would receive 2. So the selfish attitude (choosing D) of both players leads to a worse outcome than an altruistic attitude (choosing C) of both players.

In general, in a strategic game, each player has several (possibly infinitely many) strategies. A *Nash equilibrium* of such a game is a sequence of strategies, one for each player, which results in payoffs that none of them regrets. More precisely, it means that none of the players can achieve a better payoff if he is given an option to unilaterally change his strategy.[18]

In the above prisoner's dilemma game, the combination (D, D) of players' strategies is a unique Nash equilibrium. This illustrates the dilemma: each player is satisfied with his choice D, yet the outcome is worse for each of them than had they both decided to cooperate (i.e., chose C).

Many natural situations can be viewed as an instance of the prisoner's dilemma game, which explains its importance. Consider, for example, the arms race. For each country, it is beneficial not to arm itself than to arm. Yet the countries end up arming themselves. Here, not arming itself can be viewed as strategy C and arming itself as strategy D. As another example, consider two companies that produce a similar product and may choose between low (strategy C) and high (strategy D) advertisement costs. Both end up heavily advertising.

The prisoner's dilemma game can be naturally generalized to an arbitrary number of players.

Notes

[1]This appears as proof number 35 in E.S. Loomis, op. cit., pp. 49–50.

[2]F. Saidak, *A new proof of Euclid's theorem*, The American Mathematical Monthly 113(10), pp. 234–243, 2006.

[3]Eratosthenes' argument is explained in detail in N. Walkup, *Eratosthenes and the Mystery of the Stades*, Convergence, 2010, https://www.maa.org/book/export/html/116342.

[4]W. Derrick and J. Herstein, *Proof without words: Ptolemy's theorem*, The College Mathematics Journal, 43(5), p. 386, November 2012. This proof was popularized by the website of A. Bogomolny, https://www.cut-the-knot.org/.

[5]See Po-Shen Loh, *A simple proof of the quadratic formula*, arXiv:1910.06709, https://arxiv.org/abs/1910.06709, 2019.

[6]See A. Bogomolny, *Regular pentagon construction by Y. Hirano*, http://www.cut-the-knot.org/pythagoras/PentagonConstruction.shtml, 2014.

[7]To see the outcome of substituting x by $u - \frac{b}{3a}$ in the equation

$$ax^3 + bx^2 + cx + d = 0,$$

let us first simplify using formula (1) given on page 165 the first three terms constituting the left-hand side of the equation:

$$a(u - \tfrac{b}{3a})^3 = a(u^3 - 3u^2 \tfrac{b}{3a} + 3u(\tfrac{b}{3a})^2 - (\tfrac{b}{3a})^3) = au^3 - bu^2 + \tfrac{b^2}{3a}u - \tfrac{b^3}{27a^2},$$
$$b(u - \tfrac{b}{3a})^2 = b(u^2 - 2\tfrac{b}{3a}u + (\tfrac{b}{3a})^2) = bu^2 - \tfrac{2b^2}{3a}u + \tfrac{b^3}{9a^2},$$
$$c(u - \tfrac{b}{3a}) = cu - \tfrac{bc}{3a}.$$

This means that the result of the above substitution yields the equation

$$au^3 + (-\frac{b^2}{3a} + c)u + (\frac{2b^3}{27a^2} - \frac{bc}{3a} + d) = 0.$$

[8]To see the outcome of the substitution of u by $\frac{p}{y} - y$ in the equation

$$u^3 + 3pu - 2q = 0$$

we first simplify the left-hand side of equation (5), again using formula (1) given on page 165:

$$u^3 + 3pu - 2q$$
$$= (\tfrac{p}{y} - y)^3 + 3p(\tfrac{p}{y} - y) - 2q$$
$$= \tfrac{p^3}{y^3} - 3\tfrac{p^2}{y^2}y + 3\tfrac{p}{y}y^2 - y^3 + 3\tfrac{p^2}{y} - 3py - 2q$$
$$= \tfrac{p^3}{y^3} - 3\tfrac{p^2}{y} + 3py - y^3 + 3\tfrac{p^2}{y} - 3py - 2q$$
$$= -y^3 + \tfrac{p^3}{y^3} - 2q.$$

This means that the result of the substitution yields the equation

$$-y^3 + \frac{p^3}{y^3} - 2q = 0.$$

[9]Recall from Appendix 14 that

$$\zeta_1 = 1, \quad \zeta_2 = \frac{1}{2}(-1 + i\sqrt{3}), \text{ and } \zeta_3 = \frac{1}{2}(-1 - i\sqrt{3}).$$

So

$$\zeta_2\zeta_3 = \frac{1}{2}(-1 + i\sqrt{3}) \cdot \frac{1}{2}(-1 - i\sqrt{3}) = \frac{1}{4}((-1)^2 - (i\sqrt{3})^2) = \frac{1}{4}(1 + 3) = 1.$$

[10]We have, for example, (recall that $D = q^2 + p^3$)

$$\begin{aligned}
\zeta_2 \sqrt[3]{-q + \sqrt{D}} \cdot \zeta_3 \sqrt[3]{-q - \sqrt{D}} \\
= \sqrt[3]{(-q + \sqrt{D})(-q - \sqrt{D})} \\
= \sqrt[3]{q^2 - D} \\
= \sqrt[3]{-p^3} \\
= -p.
\end{aligned}$$

[11]To see the outcome of substituting x by $y - \frac{b}{4a}$ in the equation

$$ax^4 + bx^3 + cx^2 + dx + e = 0$$

we begin by simplifying the first four terms constituting the left-hand side of the equation, using formula (1) given on page 165 and the formula

$$(a + b)^4 = a^4 + 4a^3b + 6a^2b^2 + 4ab^3 + b^4.$$

$$\begin{aligned}
a(y - \tfrac{b}{4a})^4 &= a(y^4 - 4y^3\tfrac{b}{4a} + 6y^2(\tfrac{b}{4a})^2 - 4y(\tfrac{b}{4a})^3 + (\tfrac{b}{4a})^4) \\
&= ay^4 - by^3 + \tfrac{6b^2}{16a}y^2 - \tfrac{b^3}{16a^2}y + \tfrac{b^4}{256a^3}, \\
b(y - \tfrac{b}{4a})^3 &= b(y^3 - 3y^2\tfrac{b}{4a} + 3y(\tfrac{b}{4a})^2 - (\tfrac{b}{4a})^3) \\
&= by^3 - \tfrac{3b^2}{4a}y^2 + \tfrac{3b^3}{16a^2}y - \tfrac{b^4}{64a^3}, \\
c(y - \tfrac{b}{4a})^2 &= c(y^2 - 2\tfrac{b}{4a}y + (\tfrac{b}{4a})^2) \\
&= cy^2 - \tfrac{bc}{2a}y + \tfrac{b^2c}{16a^2}, \\
d(y - \tfrac{b}{4a}) &= dy - \tfrac{bd}{4a}.
\end{aligned}$$

Next, note that

$$\begin{aligned}
\frac{6b^2}{16a} - \frac{3b^2}{4a} + c &= -\frac{3b^2 - 8ac}{8a}, \\
-\frac{b^3}{16a^2} + \frac{3b^3}{16a^2} - \frac{bc}{2a} + d &= \frac{b^3 - 4abc + 8a^2d}{8a^2}, \\
\frac{b^4}{256a^3} - \frac{b^4}{64a^3} + \frac{b^2c}{16a^2} - \frac{bd}{4a} &= \frac{-3b^4 + 16ab^2c - 64a^2bd}{256a^3}.
\end{aligned}$$

This means that the result of the substitution yields the equation

$$ay^4 - \frac{3b^2 - 8ac}{8a}y^2 + \frac{b^3 - 4abc + 8a^2d}{8a^2}y + \frac{-3b^4 + 16ab^2c - 64a^2bd + 256a^3e}{256a^3} = 0.$$

^{12}Consider the equations

$$0 = s + u,$$
$$p = v + su + t,$$
$$q = tu + vs,$$
$$r = tv.$$

Then $s = -u$. This substitution leads to three equations in u, t, and v:

$$p = v - u^2 + t,$$
$$q = (t - v)u,$$
$$r = tv.$$

Isolating in the first two equations t and adding up, and subsequently, repeating the same for v, we get

$$2t = p + u^2 + \tfrac{q}{u},$$
$$2v = p + u^2 - \tfrac{q}{u}.$$

Substituting these values for t and v in the third equation, we obtain the equation

$$4r = (p + u^2 + \tfrac{q}{u})(p + u^2 - \tfrac{q}{u}),$$

which after multiplying both sides by u^2 and rearranging the terms, yields the equation

$$u^6 + 2pu^4 + (p^2 - 4r)u^2 - q^2 = 0.$$

^{13}From the equations from the previous footnote, we get

$$s = -\sqrt{u^2},$$
$$t = \frac{p + u^2 + \frac{q}{u}}{2},$$
$$v = \frac{p + u^2 - \frac{q}{u}}{2}.$$

^{14}See Chapter 5 of C. Smoryński, *History of Mathematics: A Supplement*, Springer, 2008.

^{15}The data and the outcomes are taken from M.J. Caulfield, *Apportioning representatives in the United States Congress—Jefferson's method of apportionment*, Convergence, 2010, https://www.maa.org/press/periodicals/convergence/apportioning-representatives-in-the-united-states-congress.

^{16}The cited article of Caulfield contains the results in a convenient Excel form, which makes it easy to find the right value $d = 1279$ by trial and error.

^{17}It was suggested by R. Aumann, a recipient of the Nobel Memorial Prize in Economic Sciences.

^{18}The notion of a Nash equilibrium is defined for games with an arbitrary number of players and, in a more general way, by allowing players to choose a probability distribution over their strategies (which intuitively corresponds to spreading one's bet which strategy is the best one).

Bibliography

Aaronson, S. (2023). P $\overset{?}{=}$ NP, `https://www.scottaaronson.com/papers/pnp.pdf`.

Aczel, P. (2006). *The Artist and the Mathematician* (High Stakes Publishing).

Almira, J. M. (2021). PCR tests with high sensitivity and specificity are truly trustworthy under high prevalence or medical prescription, *MATerials MATemàtics* **2021**, 5, pp. 1–5, `https://mat.uab.cat/web/matmat/wp-content/uploads/sites/23/2021/12/v2021n06.pdf`.

Almira, J. M. (2022). *Norbert Wiener: A Mathematician Among Engineers* (World Scientific).

Anon (2012a). Black–Scholes: The maths formula linked to the financial crash, BBC News, 28 April 2012, `https://www.bbc.com/news/magazine-17866646`.

Anon (2012b). Egyptian fraction, Wikipedia, `http://en.wikipedia.org/wiki/Egyptian_fraction`.

Anon (2017). Carbon dating reveals earliest origins of zero symbol, BBC News, 14 September 2017, `http://www.bbc.com/news/uk-england-oxfordshire-41265057`.

Asimov, I. (1987). *Asimov's New Guide to Science* (Penguin Books).

Aughton, P. (2011). *The Story of Astronomy* (Quercus Publishing).

Balinski, M. and Laraki, R. (2013). How best to rank wines: majority judgment, in E. Giraud-Héraud and M.-C. Pichery (eds.), *Wine Economics: Quantitative Studies and Empirical Applications* (Palgrave Macmillan, UK), pp. 149–172.

Barrow, J. D. (2008). *Cosmic Imagery: Key Images in the History of Science* (W.W. Norton & Company).

Basalla, G. (2006). *Civilized Life in the Universe: Scientists on Intelligent Extraterrestrials* (Oxford University Press).

Basu, K. (2016). A new and rather long proof of the Pythagorean theorem by way of a proposition on isosceles triangles, *The College Mathematics Journal* **47**, 5, pp. 356–360.

Belhoste, B. (1991). *Augustin-Louis Cauchy: A Biography* (Springer-Verlag).

Bell, E. T. (1965). *Men of Mathematics* (Simon & Schuster), originally published in 1937.

Bell, E. T. (1992). *The Development of Mathematics* (Dover Publications), originally published in 1940.

Berlinghoff, W. P. and Gouvêa, F. Q. (2015). *Math through the Ages: A Gentle History for Teachers and Others*, 2nd edn. (The Mathematical Association of America).

Bhattacharya, A. (2021). *The Man from the Future: The Visionary Life of John von Neumann* (Allen Lane).

Bogomolny, A. (2014). Regular pentagon construction by Y. Hirano, http://www.cut-the-knot.org/pythagoras/PentagonConstruction.shtml.

Boorstin, D. J. (1985). *The Discoverers* (Random House, Inc.), first Vintage Books Edition.

Bressoud, D. (2007). *A Radical Approach to Real Analysis* (The Mathematical Association of America).

Bronowski, J. (1973). *The Ascent of Man* (British Broadcasting Corporation).

Bronowski, J. and Mazlish, B. (1960). *The Western Intellectual Tradition* (Dorset Press).

Buchwald, J. Z. and Feingold, M. (2012). *Newton and the Origin of Civilization* (Princeton University Press).

Budiansky, S. (2021). *Journey to the Edge of Reason: The Life of Kurt Gödel* (Norton & Company).

Burton, D. (2011). *The History of Mathematics: An Introduction*, 7th edn. (McGraw-Hill Science).

Casselman, B. (2012). All for nought, American Mathematical Society, http://www.ams.org/samplings/feature-column/fcarc-india-zero.

Caulfield, M. J. (2010). Apportioning representatives in the United States Congress—Jefferson's method of apportionment, *Convergence*, https://www.maa.org/press/periodicals/convergence/apportioning-representatives-in-the-united-states-congress.

Clark, W. E. (1930). *The Aryabhatiya of Aryabhata: An Ancient Indian Work on Mathematics and Astronomy* (University of Chicago Press), republished by Kessinger Publishing, LLC in 2010.

Conan Doyle, A. (2010). *A Study in Scarlet* (CreateSpace), first published in 1887.

Connor, S. (2010). The core of truth behind Sir Isaac Newton's apple, *Independent*, 18 January 2010, http://www.independent.co.uk/news/science/the-core-of-truth-behind-sir-isaac-newtons-apple-1870915.html.

Courant, R. and Robbins, H. (2009). *What is Mathematics?* (Oxford University Press), originally published in 1941. Revised by I. Stewart.

Danson, E. (2006). *Weighing the World* (Oxford University Press).

Danzig, T. (2007). *Number: The Language of Science* (Plume), originally published in 1930.

Davis, M. (2000). *Engines of Logic* (W.W. Norton & Company).

Deakin, M. A. B. (1994). Hypathia and her mathematics, *The American Mathematical Monthly* **10**, 3, pp. 234–243.

Derrick, W. and Herstein, J. (2012). Proof without words: Ptolemy's theorem, *The College Mathematics Journal* **43**, 5, p. 386.

Devlin, K. (1988). *Mathematics: The New Golden Age* (Pelican Books).

Devlin, K. (2010). *The Unfinished Game: Pascal, Fermat, and the Seventeenth-*

Century Letter that Made the World Modern (Basic Books).

Donnelly, P. (2005). Appealing statistics, *Significance* **2**, 1, pp. 46–48, `https://doi.org/10.1111/j.1740-9713.2005.00089.x`.

Doxiadis, A. and Papadimitriou, C. (2009). *Logicomix: An Epic Search for Truth* (Bloomsbury).

Dudley, U. (1995). What to do when the trisector comes, *The Mathematical Intelligencer* **5**, 1, p. 21.

Duncan, D. E. (1998). *The Calendar* (Fourth Estate Limited).

Dunham, W. (1991). *Journey Through Genius* (Penguin).

Dunham, W. (1994). *Mathematical Universe* (John Wiley & Sons).

Dunham, W. (2005). *The Calculus Gallery* (Princeton University Press).

Dzielska, M. (1996). *Hypatia of Alexandria* (Harvard University Press), reprint Edition.

Einstein, A. (2015). *The Theory of Relativity and Other Essays* (Philosophical Library/Open Road).

Euclid (2002). *Euclid's Elements* (Green Lion Press), translated by T.L. Heath in 1908, edited by D. Densmore.

Fancher, R. F. (1996). *Pioneers in Psychology*, 3rd edn. (W.W. Norton & Company).

Farmelo, G. (2019). *The Universe Speaks in Numbers* (Faber & Faber Ltd).

Feferman, A. B. and Feferman, S. (2004). *Alfred Tarski: Life and Logic* (Cambridge University Press).

Fomenko, A. T. (2007). *History: Fiction Or Science? Chronology. Vol. I*, second, revised and expanded edn. (Delamere Publishing).

Fortnow, L. (2013). *The Golden Ticket: P, NP, and the Search for the Impossible* (Princeton University Press).

Freely, J. (2009). *Alladin's Lamp* (Alfred A. Knopf).

Frege, G. (2007). *The Foundations of Arithmetic, Volume II* (Routledge), originally published in 1903.

Galilei, G. (2010). *Dialogues Concerning Two New Sciences* (Cosimo Classics), originally published in 1638.

Gamut, L. T. F. (1991). *Introduction to Logic* (The University of Chicago Press).

Goldstein, R. (2009). *Betraying Spinoza: The Renegade Jew Who Gave Us Modernity* (Schocken).

Gorroochurn, P. (2016). *Classic Topics on the History of Modern Mathematical Statistics* (Wiley).

Gottlieb, A. (2000). *The Dream of Reason* (Allen Lane).

Gottlieb, A. (2016). *The Dream of Enlightenment* (Allen Lane).

Gowers, T. (2012). *Mathematics: A Very Short Introduction* (Oxford University Press).

Gowers, T., Leader, I., and Barrow-Green, J. (eds.) (2008). *The Princeton Companion to Mathematics* (Princeton University Press).

Grattan-Guinness, I. (2000). *The Rainbow of Mathematics* (W.W. Norton & Company).

Gray, J. (2016). "Let us calculate!": Leibniz, Llull, and the computational imagination, *The Public Domain Review*, `http://publicdomainreview.org/2016/11/10/let-us-calculate-`

leibniz-llull-and-computational-imagination/.

Grimes, W. (2011). Roman Opalka, an artist of numbers, is dead at 79, *New York Times*, 9 August 2011, http://www.nytimes.com/2011/08/10/arts/design/roman-opalka-conceptual-artist-with-numerical-focus-is-dead-at-79.html.

Hald, A. (2003). *A History of Probability and Statistics and Their Applications before 1750* (Wiley-Interscience).

Hartshorne, R. (2000). *Geometry: Euclid and Beyond* (Springer).

Heaton, L. (2015). *A Brief History of Mathematical Thought* (Robinson).

Heilbron, J. L. (2000). *Geometry Civilized: History, Culture, and Techniques* (Oxford University Press).

Hellman, H. (2006). *Great Feuds in Mathematics* (Wiley).

Hirshfeld, A. W. (2001). *Parallax: The Race to Measure the Cosmos* (W. H. Freeman and Company).

Hoffman, P. (1998). *The Man Who Loved Only Numbers: The Story of Paul Erdős and the Search for Mathematical Truth* (Hachette Books).

Hogben, L. (1940). *Mathematics for the Million* (George Allen & Unwin Ltd).

Hollingdale, S. (2011). *The Makers of Mathematics* (Dover Publications).

Isaacson, W. (2017). *Leonardo da Vinci* (Simon & Schuster).

Jackson, A. (2004a). Comme appelé du néant—as if summoned from the void: the life of Alexandre Grothendieck I, *Notices of the AMS* **51**, 4, pp. 1038–1056, https://www.ams.org/notices/200409/fea-grothendieck-part1.pdf.

Jackson, A. (2004b). Comme appelé du néant—as if summoned from the void: the life of Alexandre Grothendieck II, *Notices of the AMS* **51**, 10, pp. 1196–1212, https://www.ams.org/notices/200410/fea-grothendieck-part2.pdf.

Joseph, G. G. (2011). *The Crest of the Peacock: Non-European Roots of Mathematics*, 3rd edn. (Princeton University Press).

Kahn, D. (1996). *The Codebreakers: The Comprehensive History of Secret Communication from Ancient Times to the Internet* (Simon & Schuster).

Karam, R. (2018). Fresnel's original interpretation of complex numbers in 19th century optics, *American Journal of Physics* **86**, 4, pp. 245–249.

Katz, V. J. (2009). *A History of Mathematics*, 3rd edn. (Addison Wesley).

Katz, V. J. and Parshall, K. H. (2014). *Taming the Unknown: A History of Algebra from Antiquity to the Early Twentieth Century* (Princeton University Press).

Keestra, M. (ed.) (2016). *Een cultuurgeschiedenis van de wiskunde* (Nieuwezijds), https://www.nieuwezijds.nl/Boek/9789057121364/Een-cultuurgeschiedenis-van-de-wiskunde/. In Dutch.

Kline, M. (1959). *Mathematics and the Physical World* (Ty Crowell Co.), republished by Dover Publications in 1981.

Kline, M. (1972). *Mathematical Thought From Ancient to Modern Times* (Oxford University Press).

Knuth, D. E. (2016). *Art of Computer Programming, Volume 4, Fascicle 4: Generating All Trees—History of Combinatorial Generation* (Addison-Wesley).

Kordos, M. (2010). *Wykłady z Historii Matematyki*, 3rd edn. (Script), in Polish.

Lerner, R. E., Meacham, S., and Burns, E. M. (1998). *Western Civilizations*, 13th edn. (W.W. Norton & Company).

Livio, M. (2002). *The Golden Ratio* (Broadway Books).

Livio, M. (2005). *The Equation That Couldn't be Solved* (Simon & Schuster).

Loomis, E. S. (1940). *The Pythagorean Proposition*, 2nd edn. (National Council of Teachers of Mathematics), available from http://www.eric.ed.gov/PDFS/ED037335.pdf.

Lui, S. H. (1997). An interview with Vladimir Arnol'd, *Notices of the AMS* **44**, 4, pp. 432–438.

Mackenzie, D. (2001). Wavelets: Seeing the forest and the trees, National Academy of Sciences, http://www.nasonline.org/publications/beyond-discovery/wavelets.pdf.

Maor, E. (2009). *e: The Story of a Number* (Princeton University Press).

Merzbach, U. C. and Boyer, C. B. (2011). *A History of Mathematics*, 3rd edn. (John Wiley & Sons).

Munro, A. (2009). *Too Much Happiness* (Douglas Gibson Books).

Nasar, S. (2011). *A Beautiful Mind* (Simon & Schuster Paperbacks), originally published in 1998.

Nasar, S. and Gruber, D. (2006). Manifold destiny. a legendary problem and the battle over who solved it, *New Yorker,* 28 August 2006, https://www.newyorker.com/magazine/2006/08/28/manifold-destiny.

Nazaryan, A. (2013). A most profound math problem, *New Yorker,* 2 May 2013, http://www.newyorker.com/tech/elements/a-most-profound-math-problem.

Netz, R. and Noel, W. (2007). *The Archimedes Codex* (Weidenfeld & Nicholson).

Netz, R., Noel, W., Wilson, N., and Tchernetska, N. (eds.) (2011). *The Archimedes Palimpsest* (Cambridge University Press).

O'Connor, J. J. and Robertson, E. F. (2000). An overview of Babylonian mathematics, http://www-history.mcs.st-and.ac.uk/HistTopics/Babylonian_mathematics.html.

O'Connor, J. J. and Robertson, E. F. (2003). Qin Jiushao, https://mathshistory.st-andrews.ac.uk/Biographies/Qin_Jiushao.

Odifreddi, P. (2004). *The Mathematical Century: The 30 Greatest Problems of the Last 100 Years* (Princeton University Press).

Odlyzko, A. M. (1995). Tragic loss or good riddance? The impending demise of traditional scholarly journals, *International Journal of Human-Computer Studies* **42**, 1, pp. 71–122, doi:https://doi.org/10.1006/ijhc.1995.1004.

Ornes, S. (2015). Researchers race to rescue the enormous theorem before its giant proof vanishes, *Scientific American,* 1 July 2015, https://www.scientificamerican.com/article/researchers-race-to-rescue-the-enormous-theorem-before-its-giant-proof-vanishes. Originally published as "The Whole Universe Catalog", *Scientific American* **313**, 1, pp. 68–75 (July 2015).

Pais, A. (1982). *Subtle Is the Lord: The Science and the Life of Albert Einstein* (Oxford University Press).

Pickover, C. A. (2009). *The Math Book* (Sterling).

Po-Shen Loh (2019). A simple proof of the quadratic formula, https://arxiv.org/abs/1910.06709.

Poundstone, W. (2006). *Fortune's Formula* (Hill and Wang).

Poundstone, W. (2009). *Gaming the Vote: Why Elections Aren't Fair (and What We Can Do About It)* (Hill and Wang).
Rittaud, B. and Heeffer, A. (2014). The pigeonhole principle, two centuries before Dirichlet, *The Mathematical Intelligencer* **36**, 2, pp. 27–29.
Robinson, A. (ed.) (2012). *The Scientists* (Thames and Hudson).
Rose, N. J. (1988). *Mathematical Maxims and Minims* (Rome Press, Inc.).
Rosenhouse, J. (2009). *The Monty Hall Problem: The Remarkable Story of Math's Most Contentious Brain Teaser* (Oxford University Press).
Roy, R. (2011). *Sources in the Development of Mathematics Series and Products from the Fifteenth to the Twenty-first Century* (Cambridge University Press).
Rupert, W. W. (2008). *Famous Geometrical Theorems and Problems* (BiblioBazaar), first published in 1900.
Russell, B. (1946). *History of Western Philosophy* (Unwin Books).
Russell, B. (1957). *Mysticism and Logic* (Doubleday Anchor Books).
Russell, B. (1975). *The Autobiography of Bertrand Russell* (Unwin Books), first published in 1967.
Rząźewski, K., Słomczyński, W., and Życzkowski, K. (2014). *Każdy Głos się Liczy!* (Wydawnictwo Sejmowe), in Polish.
Sabra, A. I. (1981). *Theories of Light: From Descartes to Newton* (Cambridge University Press).
Saidak, F. (2006). A new proof of Euclid's theorem, *The American Mathematical Monthly* **113**, 10, pp. 937–938.
Saitoh, S. (2021). History of division by zero and division by zero calculus, *International Journal of Division by Zero Calculus* **1**, 1, pp. 1–38.
Salsburg, D. (2001). *The Lady Tasting Tea: How Statistics Revolutionized Science in the Twentieth Century* (W.H. Freeman and Company).
Seife, C. (2000). *Zero: The Biography of a Dangerous Idea* (Penguin).
Sesiano, J. (1982). *Books IV to VII of Diophantus' Arithmetica in the Arabic Translation Attributed to Qusta ibn Luqa* (Springer-Verlag).
Shenitzer, A. and Stillwell, J. (eds.) (2002). *Mathematical Evolutions* (The Mathematical Association of America).
Simmons, G. F. (2007). *Calculus Gems, Brief Lives and Memorable Mathematics* (The Mathematical Association of America).
Singh, S. (1998). *Fermat's Enigma: The Epic Quest to Solve the World's Greatest Mathematical Problem* (Anchor).
Smoryński, C. (2008). *History of Mathematics: A Supplement* (Springer).
Sparavigna, A. (2013). The science of Al-Biruni, *International Journal of Sciences* **2**, 12, pp. 52–60.
Steinhaus, H. and Trybula, S. (1959). On a paradox in applied probabilities, *Bulletin of the Polish Academy of Sciences* **7**, pp. 67–69.
Stewart, I. (2008). *Taming the Infinite* (Quercus Publishing).
Stewart, I. (2012). *In Pursuit of the Unknown: 17 Equations That Changed the World* (Basic Books).
Stewart, I. (2014). *Great Problems of Mathematics* (Profile Books Ltd).
Stewart, I. (2017a). *Infinity: a Very Short Introduction* (Oxford University Press).
Stewart, I. (2017b). *Significant Figures: The Lives and Work of Great Mathematicians*

(Basic Books).

Stigler, S. M. (1986). *The History of Statistics: The Measurement of Uncertainty before 1900* (Belknap Press).

Stigler, S. M. (2016). *The Seven Pillars of Statistical Wisdom* (Harvard University Press).

Stillwell, J. (2010). *Mathematics and its History*, 3rd edn. (Springer).

Stillwell, J. (2016). *Elements of Mathematics: From Euclid to Gödel* (Princeton University Press).

Strickland, L. and Lewis, H. (2022). *Leibniz on Binary: The Invention of Computer Arithmetic* (The MIT Press).

Struik, D. J. (1987). *A Concise History of Mathematics*, 4th edn. (Dover Publications).

Strzelecki, P. (2011). *Matematyka Współczesna dla Myślących Laików* (Wydawnictwa Uniwersytetu Warszawskiego), in Polish.

Stumpf, S. E. (1989). *Philosophy: History and Problems* (McGraw-Hill Book Company).

Szpiro, G. G. (2008). *Poincaré's Conjecture* (A Plume Book).

Szymborska, W. (2000). *Poems New and Collected* (Mariner Books), translated from Polish by S. Barańczak and C. Cavanagh.

Tabak, J. (2004a). *Algebra: Sets, Symbols, and the Language of Thought*, The History of Mathematics (Facts on File).

Tabak, J. (2004b). *Geometry: The Language of Space and Form*, The History of Mathematics (Facts on File).

Tabak, J. (2004c). *Numbers: Computers, Philosophers, and the Search for Meaning*, The History of Mathematics (Facts on File).

Tabak, J. (2004d). *Probability and Statistics: The Science of Uncertainty*, The History of Mathematics (Facts on File).

Tegmark, M. (2014). *Our Mathematical Universe* (Alfred A. Knopf).

Thompson, C. (2007). Death of checkers, *New York Times*, 9 December 2007, http://www.nytimes.com/2007/12/09/magazine/09_15_checkers.html.

Tierney, J. (1991). Behind Monty Hall's doors: puzzle, debate and answer? *New York Times*, 21 July 1991, http://www.nytimes.com/1991/07/21/us/behind-monty-hall-s-doors-puzzle-debate-and-answer.html.

Tijms, H. (2021a). *Basic Probability: What Every Math Student Should Know*, 2nd edn. (World Scientific).

Tijms, H. (2021b). *Chance, Logic and Intuition: An Introduction to the Counter-Intuitive Logic of Chance* (World Scientific).

van Heijenoort, J. (ed.) (1967). *From Frege to Gödel* (Harvard University Press).

Vitányi, P. M. B. (2007). Andrey Nikolaevich Kolmogorov, *Scholarpedia* 2, 2, p. 2798, http://www.scholarpedia.org/article/Andrey_Nikolaevich_Kolmogorov.

Walkup, N. (2010). Eratosthenes and the mystery of the stades, *Convergence*, https://www.maa.org/book/export/html/116342.

Watson, P. (2005). *Ideas: A History of Thought and Invention, from Fire to Freud* (Weidenfeld & Nicholson).

Weyl, H. (1952). *Symmetry* (Princeton University Press).

Wigner, E. (1960). The unreasonable effectiveness of mathematics in the natural sciences, *Communications in Pure and Applied Mathematics* **13**, pp. 1–14.

Wolchover, N. (2013). Dispute over infinity divides mathematicians, *Scientific American*, 3 December 2013. Originally published in *Quanta Magazine*, https://www.scientificamerican.com/article/infinity-logic-law/.

Yandell, B. (2001). *The Honors Class: Hilbert's Problems and Their Solvers* (A K Peters/CRC Press).

Young, P. (1995). Optimal voting rules, *The Journal of Economic Perspectives* **9**, 1, pp. 51–64.

Name Index

Galileo Galilei, 53, 61, 62, 65, 71, 74n, 75n, 84, 105, 109n, 147
Galois, É., 95–97, 100, 104, 122
Galton, F., 100, 101, 118
Gamut, L.T.F., 26n
Garfield, J.A., 10, 86
Gauss, C.-F., 73n, 92–96, 98, 99, 107, 109n, 136
Gibbon, E., 69
Giraud-Héraud, E., 88n
Goldstein, R., 27n
Gorroochurn, P., 108n
Gottlieb, A., 26n, 75n
Gouvêa, F.Q., xn, 74n, 152
Gowers, T., xn, 109n, 140, 143, 144n, 154
Gödel, K., 114–116, 122, 126, 131, 137, 149
Grattan-Guinness, I., 46, 49n, 74n, 75n, 153
Gray, J., 48n, 75n
Gregory, J., 75n
Gregory XIII, 37
Grimes, W., 48n
Grothendieck, A., 138, 144n
Gruber, D., 144n

Hadamard, J., 96, 107, 110n
Haken, W., 135
Hald, A., 75n
Hales, Th., 136
Hamming, R., 123
Hammurabi, 2
Hardy, G.H., 137, 144n
Hartshorne, R., 16, 27n
Heaviside, O., 109n
Heeffer, A., 109n
Heilbron, J.L., 27n
Hellman, H., 75n, 88n
Henri III, King of France, 57
Henri IV, King of France, 57
Heron of Alexandria, 21, 48n, 51
Herstein, J., 161, 193n
Hilbert, D., ix, xn, 27n, 111–115, 117, 119, 140, 146, 149
Hipparchus of Nicaea, 18, 19, 22

Hirano, Y., 168
Hirshfeld, A.W., 27n, 74n
Hoare, C.A.R., 131
Hobbes, Th., 15, 18, 68
Hodges, A., 142n, 154
Hoffman, P., 144n
Hogben, L., 74n
Hollingdale, S., 87n
Homer, 10
Howard, R., 154
Huygens, Ch., 69, 70, 72, 79, 80
Hypatia of Alexandria, 23, 27n, 145

Infeld, L., 154
Isaacson, W., 27n

Jackson, A., 144n
Jefferson, Th., 72, 85, 86, 181
Joseph, G.G., 7n, 35, 36n, 44, 48n, 73n
Joyce, D.E., 26n
Julius Caesar, Roman Emperor, 7, 37, 38, 46

Kahn, D., 48n
Kanigel, R., 144n, 154
Kant, I., 90, 112, 148
Karam, R., 73n
Kasparov, G., 143n
Katz, V.J., 8n, 48n, 109n, 153
Kehlmann, D., 154
Kepler, J., 48n, 60, 61, 63, 65, 71, 74n, 136
Keynes, J.M., 71, 75n
Kline, M., 66, 73–75n, 111, 142n, 153
Knuth, D.E., 49n, 133, 140
Kolmogorov, A.N., 117, 122, 149
Kovalevsky, S., 145, 146
Kulik, J.P., 42
Kuratowski, K., 129, 130

Lagrange, J.-L., 41, 89–92, 96, 99
Lambert, J.H., 26n
Laplace, de P.-S., 86, 90–92, 99, 106
Laraki, R., 88n
Leader, I., xn, 109n, 143n, 154
Leavitt, D., 154

Ptolemy, Claudius, 18, 21, 39, 58, 59
Pythagoras of Samos, 3, 10n, 10, 11n, 14n, 15, 23n, 29, 31, 33, 62, 136, 156, 158, 169, 174

Qin Jiushao, 43
Quetelet, A., 99, 100

Ramanujan, S., 137, 154
Ramsey, F.P., 129, 143n
Raphael, 15
Regiomontanus, 56
Rhind, H., 4
Riemann, B., 98, 112, 123, 124
Rittaud, B., 109n
Rivest, R., 17
Robbins, H., 153
Robertson, E.F., 7n, 48n
Robertson, N., 130
Roberval, de G., 75n
Robinson, A., 75n, 124
Rooney, A., 152
Rose, N.J., 143n
Rosenhouse, J., 87n
Ruffini, P., 94, 95
Russell, B., 10, 15, 23, 26n, 27n, 105, 113, 117, 119, 137, 142n, 148, 149, 150n

Sabra, A.I., 75n
Saidak, F., 160, 193n
Saitoh, S., 48n
Sartre, J.-P., 139
Savant, vos M., 80
Schaeffer, J., 143n
Schogt, P., 154
Scholes, M., 128, 143n
Sesiano, J., 48n
Seymour, P., 130
Shamir, A., 17
Shanks, W., 13
Shannon, C., 118, 119
Shenitzer, A., 143n
Siegmund-Schultze, R., 143n
Simmons, G.F., 75n, 108n
Singh, S., 144n

Smith, A., 127
Smoryński, C., 174, 195n
Snel van Rooyen, W., *see* Snellius, W.
Snellius, W., 9, 26n
Socrates, 13, 15
Sparavigna, A., 48n
Spinoza, B., 15, 27n
Steinhaus, H., 142n
Stevin, S., 56, 74n
Stewart, I., xn, 109n, 143n, 152, 153
Stigler, S.M., 109n
Stillwell, J., 27n, 143n, 154
Strickland, L., 75n
Struik, D.J., 153
Strzelecki, P., 144n
Stumpf, S.E., 26n
Sylvester, J.J., 102, 103
Szymborska, W., 12

Tabak, J., 152, 153
Tao, T., 140
Tarski, A., 115, 116
Tartaglia, N., 53, 54, 171
Tchernetska, N., 27n
Tegmark, M., 150n
Thales of Miletus, 9, 33
Thompson, C., 143n
Thurston, W., 139
Tierney, J., 87n
Tijms, H., 109n
Tijms, S., 87n
Torricelli, E., 75n
Trybula, S., 142n
Turing, A., 114, 116, 117, 131, 137, 154
Twain, M., 99
Tyldum, M., 154

Valéry, P., 25
Vallée-Poussin, de la Ch.J., 108
van Ceulen, L., 12, 83
van Heijenoort, J., xn
Vasari, G., 51
Viazovska, M.S., 146
Viète, F., 56, 57, 171
Virgil, 81